VGM CAREER PLANNER SERIES

High Tech

A VGM Career Planner

Gary D. Golter

Deborah F. Yanuck

Foreword by
Walter K. Weisel
President and COO
Prab Robots, Inc.

VGM Career Horizons
a division of *NTC Publishing Group*
Lincolnwood, Illinois USA

Library of Congress Cataloging-in-Publication Data

Golter, Gary D.
  High tech.

  (VGM career planner series)
  Previously published as: Opportunities in high tech careers. ©1987.
  1. High technology industries—Vocational guidance.
I. Yanuck, Deborah F.   II. Title.   III. Series.
T49.5.G65   1988   620   88-61874
ISBN 0-8442-8681-8

Published by VGM Career Horizons, a division of NTC Publishing Group.
© 1989 by NTC Publishing Group, 4255 West Touhy Avenue,
Lincolnwood (Chicago), Illinois 60646-1975 U.S.A.
All rights reserved. No part of this book may be reproduced, stored
in a retrieval system, or transmitted in any form or by any means,
electronic, mechanical, photocopying, recording or otherwise, without
the prior permission of NTC Publishing Group.
Manufactured in the United States of America.

8 9 0 ML 9 8 7 6 5 4 3 2 1

About the Authors

Gary D. Golter has been in the computer software business for 28 years. He is president of Computer Business Consultants in Woodland Hills, California, a firm specializing in the creation and distribution of expert systems that address the reallocation of human resources. He is a pioneer in the application of the use of computers in the rehabilitation of disabled people. He oversees the development of software, systems design, and marketing. He has also had several books published on the use of computers in private sector rehabilitation.

Mr. Golter received his bachelor of arts degree from Cal State University, Northridge, California. Mr. Golter lives in southern California with his wife, Marjorie, and his six children.

Deborah F. Yanuck received her bachelor of arts degree from University of California, Los Angeles (UCLA). She has worked as a writer and research assistant for Computer Business Consultants doing job analysis, and analyzing economic and employment opportunities in high tech jobs. Previously, she has worked for other authors including Freda Greene, author of *How to Get A Job in Los Angeles,* doing occupational research and editing text.

Ms. Yanuck lives in southern California with her husband, Andy, and her daughter, Danielle.

Acknowledgments

We would like to thank the people who used their unique expertise and knowledge of the high technology industry to edit and provide hard-to-get information for this book. These people are Michael Mulchahy of RCA, Thomas Murname of Southern California Edison Company, Richard Powers of Rocketdyne, and J. Harold Saks. We appreciate all the time they spent improving our book.

GARY GOLTER AND DEBORAH YANUCK

Foreword

There is a saying among business people and educators that one's education begins when one graduates. In this era of rapid change in job opportunities, the saying is more true than ever. Since the industrial revolution the work place has changed at an accelerating pace. With the advent of the computer and all the innovations it has spawned, change permeates virtually every occupation.

One might characterize the work place of today as one of occupational obsolescence. Jobs which were considered secure only a decade or so ago are today being phased out or, at least, drastically changed. The half-life of an engineer for example has steadily declined from ten years to less than five years. People who manage businesses require information much faster and rely on data provided by the computer.

The computer has become ingrained in business and it requires constant attention. Computers have become as much a part of the home as the television and radio. The question is not whether or not to buy one; the question is how to utilize it most effectively.

Today's employees and the employees of the future will have to grasp the importance of this new tool. Virtually every job of the future will in one way or another be affected by the computer. To remain competitive in the job market one will have to understand its use. This will require changing attitudes about the job market.

Those of us who work in high tech fields can appreciate that such integration of new and evolving technologies means constant training and retraining. It will require spending more time in the classroom, studying at home and during one's leisure, and seeking every opportunity to upgrade one's saleable skills.

Perhaps one of the most difficult jobs facing all industries today is to provide retraining for their workers. Most jobs now require the use of such high technology education devices as computers, robots, numerically controlled machine tools, and factory monitoring systems. Those willing to acknowledge the changing environment and sharpen their skills through new training will be the leaders and supervisors of tomorrow's working generation.

Career paths for graduates most definitely should provide for alternate opportunities. Even individuals already employed should seriously look at the opportunities inside and outside their present job or profession. Will your present job exist ten years from now?

This book is intended to help direct individuals in evaluating the scope and direction of their chosen occupations. It explores the areas of high technology and suggests some alternative avenues of career pathing. While this book cannot predict the future, it acknowledges that changing economic conditions, the impact of high technology, and the need for realistic job market appraisal must become part of every individual's employment portfolio.

One should not fear change or high technology. Instead, being prepared for change may be your employment passport to the high technology future.

Walter K. Weisel
President and COO
Prab Robots, Inc.

Contents

About the Authors iii

Acknowledgments v

Foreword vii

1. **High Tech Careers: An Overview** 1

2. **Robotics** 7
   What is robotics? Careers in robotics. Robotics engineer. Robotic technologist/technician. Robotic maintenance/servicer person. Robot sales representative. Robot educator/trainer. Summary.

3. **Space Technology** 15
   The space technology industry. Careers in space technology. Engineering-development technician. Engineering-document-control clerk/configuration management. Guidance navigation and control engineer. Guidance and control electronics/microprocessor engineer. Satellite systems engineer. Control system analyst. Real-time simulation programmer. Performance analysis engineer. Satellite attitude estimation and control analyst. Laser communication engineer. Optical and electro-optical systems analysis engineer. Engineering systems analyst. Image processing analyst. Scientific programmer/analyst. Software engineer. Other career opportunities.

4. Telecommunications **29**

The telecommunications industry. Careers in telecommunications. Account executive. Technical consultant. Chemist and chemical engineer. Materials engineer. Physicist. Electro-optics engineer. Electrical engineer. Electronic engineer. Computer scientist. Mechanical engineer. Industrial engineer. Summary.

5. Biotechnology **43**

The biotechnology industry. Careers in biotechnology. Associate scientist/scientist. Senior biostatistician. Scientific programmer. Laboratory associate/research assistant. Technical secretary. Data entry assistant. Senior instrumentation technician. Senior maintenance mechanic. Laboratory head. Research scientist. Research associate. Nuclear medicine technologist. Summary.

6. Computer Industry **55**

Computer technology. Careers in the computer industry. Designer. Engineer. Project manager. Manager of research & development. Independent consultant. Technician. Fabricator. Line supervisor. Technical writer. Course instructor. Education and training administrator. Software designer. Systems analyst. Manager of planning. Documentation librarian. Business applications programmer. Scientific applications programmer. Contract programmer. Data-base administrator. Telecommunications network analyst. Chief programmer. Manager-systems development. Support or maintenance programmer. Systems programmer. Manager-system support. Computer service technician. Computer equipment technical specialist. Service manager. Sales representative. Sales/marketing engineer. Sales/marketing manager.

7. Energy Industry **77**

The energy industry today. Careers in the energy industry. Maintenance/operations positions. Administrative/professional positions. Skilled/journeyman positions.

Appendix: Résumés, Application Forms, Cover Letters, and Interviews **97**

CHAPTER ONE

HIGH TECH CAREERS: AN OVERVIEW

THIS BOOK IS FOR THE STUDENT, THE DISPLACED WORKER, THE corporate executive and anyone interested in today's labor market. It is a guide to the new world of employment. It provides and explains:

- what work opportunities are out there
- how to decide which field of work meets your qualifications
- how to prepare for new high technology jobs
- information on training and education required for available jobs
- what realistic jobs to plan for from now until the year 2000
- information to help everyone prepare for the present and future job market
- information that helps in searching for a job
- illustrations of trends in the job market

This book explores the emergence of new high technology industries and how you, the worker, student, or executive can become a successful part of this exciting revolution. You will learn what training and education are needed to obtain employment in high technology industries and what skills can be transferred from past jobs to today's and tomorrow's labor market.

Many people have heard of the new high technology industries, but tend to think of them in abstract terms, not having a personal effect. If you are currently employed, you cannot ignore the fact that you might need some retraining to continue working in the changing job market. If you are a student, you must consider colleges and vocational schools that are prepared to offer training in new technologies. If you are a corporate executive in one of the old line industries, you need to start planning the retraining of your managers and workers so they can adapt to this changing industrial environment.

The Labor Department estimates that, from 1979 through 1983, about 5 million workers lost their jobs because of plant closings. Only about 3 million of them eventually found new jobs. People who have been assembly line workers find it difficult to adapt to jobs in new high tech areas. There are many reasons for this difficulty. These people often do not have the skills or training necessary for the new jobs and the jobs usually pay less than the worker's previous union job. Also, people do not like to relocate in search of employment unless they can be assured of finding jobs. Unfortunately, displaced workers who are unable or unwilling to be retrained or relocated may never reenter the workforce.

While advances in technology create new jobs for skilled and highly trained workers, there will be millions of workers displaced by changes in the economy by the year 2000. Even though many of these workers have skills that can be used in other jobs, the worker usually does not realize that he or she has these transferable skills. A case illustrating this is described below:

> A man had a job wiring and repairing telephones that ended in a permanent layoff. He felt that he was not skilled for anything other than telephone work and that he had wasted all those working years. In this case, the vocational counselor who reviewed his previous job duties recognized, that as part of his duties, the man had also

processed billing and invoices using a computer terminal. It was therefore determined by the counselor that this man was well equipped to switch to white collar clerical work, an area where many new jobs are found today.

This example illustrates how important information about past work skills is in finding a new job. Without such information, the counselor would have been unable to help this man.

New jobs emerging in today's labor market exemplify the nation's shift from heavy industry to high technology, information, and service industries. No longer will there be a wide availability of jobs in automobile and steel plants. Many of these plants are closing their doors or converting from labor intensive assembly lines to automation and the use of robots. The use of robots and the development of other new technologies is creating more and more technically based job opportunities as unskilled jobs are phased out.

The high technology sector has contributed significantly to economic expansion, employment opportunities, and national productivity. Although studies confirm that the high tech industries offered rapid and significant job growth prospects even during the recession of the 1980s, these gains do not appear to have compensated completely for job losses in other sectors. For example, estimates indicate that the workforce in robotics will grow to 100,000 by 1990 through new jobs in installation, maintenance, and servicing. Adding the fields of biogenetics, software, and photovoltaics, approximately 500,000 new jobs will be available by 1990. This figure represents more than double the present jobs in these areas. If we include related support jobs generated by these new fields, it is estimated that jobs created in these four high tech fields will reach nearly three million by 1990; however is it also expected that 25 million jobs will be eliminated by high technology. It is difficult to determine whether the new expanding technology will result in an overall net gain or net loss of jobs. All these figures and estimates must be balanced with the creation of other new jobs in the many other new high tech industries. We explore this problem and discuss ways of identifying industries with new employment opportunities in subsequent chapters. The education and training that will be required is also examined. The displaced worker shall be able to assess his or her present job skills and see how he or she can use these skills in new high tech pursuits.

Many jobs in high tech sectors are, and will remain, clerical and service jobs. The need for professional and skilled blue collar jobs will be much less. So it appears that high technology fields will exacerbate the loss of middle jobs. There will continue to be a small number of professional workers experiencing higher incomes and challenging work, and a large number of blue collar workers displaced by the new technologies. Many new jobs created will be in lower paid clerical, service, and assembly occupations.

There is some evidence that existing high tech geographical centers will continue to dominate as headquarters for these new high technology firms. At present, plants and jobs are highly concentrated in a very few states. For example, half of all biogenetics firms are located in California and Massachusetts. Two factors explain the tendency for these fields to concentrate in a few places. First, high technology firms tend to remain in the location where they were developed. For example, production of robotics occurs in the traditional machining sector, that is, older manufacturing centers like Detroit. Second, high tech sectors require a pool of highly skilled professional and technical workers, usually available where good universities are located.

High tech firms tend to be very conscious of the prestige attached to the area in which they are located. For this reason high tech firms generally do not exist in innercity areas or areas that have lower labor costs and inexpensive rents. Therefore new high tech jobs will not be available in all geographical areas; inner cities and certain other regions will remain excluded from this new source of job expansion. Along this same line, communities such as Austin, Texas, which is experiencing industrial growth, continue growing and are becoming overcrowded.

This book can be used by students as a guide to the general educational preparation necessary for a job that will actually exist and be in demand when they graduate. Upon graduating from school many students feel as if their education was a waste of time and that they have not learned any skills or gained any knowledge that will enable them successfully to find a job. This attitude is often accompanied by uncertainty. The student does not know what career he or she wants to enter. Because of this lack of direction, many students float through school taking the easiest classes needed to graduate and then leave with no applicable education or training.

Using the information in this book, students will be able to prepare themselves for jobs that are in demand and emerging in the high technology industry by learning what courses they must take during their high school and college years. For example, since almost all jobs now require some work with computers, it is very important that students understand and know how to use computers.

One of the jobs most in demand now is that of the *engineer*. The student who earns an engineering or computer science degree is best prepared for the new high technology jobs. Engineers will design, produce, and implement the new technologies associated with robotics, new telecommunications systems, and other high technology products. However, excellent job opportunities will still be found in marketing, customer relations, personnel and other service related occupations.

We know that this book will enable you to become an integral part in the development and expansion of the new high technology, information, and service based economy.

CHAPTER TWO

ROBOTICS

What is Robotics?

The Robotic Industries Association (RIA) defines *ROBOTICS* as "a reprogrammable multifunctional manipulator designed to move material, parts, tools, or specialized devices through variable programmed motions for the performance of a variety of tasks." All sectors of the economy are looking to these highly complex programmable machine tools for productivity improvement, uniformity, safety, and cost reduction.

In the robotics industry, sales and production are expanding at the impressive average annual rate of 35-40 percent. In 1981 robot sales by American manufacturers exceeded 150 million dollars, more than a 60% increase over 1980. This increase will most likely level out to 35-40%, but either figure accurately demonstrates the continuing expansion of robotics.

There appear to be three primary positive aspects to this emerging industry. First, again quoting a spokesperson of the RIA, "Robot capabilities will steadily increase and diversify, and those improving qualities will enable the gradual incorporation of robots into Flexible Manufacturing Systems (FMS) that will combine various forms of factory automation into fully mechanized and self-sufficient production processes." Second, engineers and specially trained electronics technicians will find an entirely new and expanding sector of employment. W. E. Upjohn Institute for Employment Research reports that robotics has the potential to create 70,000 new jobs by 1990, with half of these jobs available for engineers and engineering technicians. As John W. Wright reports, "The demand basically is for two types of technically trained people. Most of all, the industry needs engineers, mechanical and electrical engineers.... Companies need engineers to design and construct robots—and to train customers [users] in ways to integrate robots into their manufacturing procedures." Third, robots in today's industries can perform work tasks that are too tedious, repetitive, hazardous, and strenuous for humans.

The expanding robotics industry has negative aspects, too. One problem is that the first group of people to lose their jobs due to robots are unskilled factory workers. Some estimates indicate that between 100,000 and 200,000 jobs will be eliminated by 1990. For example, the jobs of welder, painter, and assembler in heavy industries will be the first to become obsolete. The Rand Corporation predicts that only two percent of the workforce will be employed in manufacturing by the year 2000, as opposed to 20 percent today. Research in Germany indicates that robots will eliminate 5 jobs for every one the new industry creates. However, many of the robotics industry's spokespersons say that displaced workers can be retrained for employment in a robot dominated manufacturing environment. They also say that robots enable companies to increase production and distribute better products at lower costs, thus expanding and increasing the overall workforce.

The emergence of the robotics industry contains the positive benefits of eliminating hazardous and tedious jobs, improving the quality and uniformity of goods, allowing around-the-clock production capabilities (since robots do not sleep), and creating more jobs for the trained technical worker. It must be kept in mind that the positive and negative aspects of the new robotics industry are only predictions.

Careers in Robotics

Many robotics experts report that it is best to start planning for a career in robotics in the ninth or tenth grade. Students interested in robotics should take courses in math, physics, chemistry, the industrial arts, social science, English, and speech. It is also important for the student planning a robotics career to possess mechanical aptitude. Working with mechanical or electrical devices or building models provides good practice for strengthening mechanical abilities. The future robotics engineer or technician should have a mind that questions everything and should be able to analyze questions in an orderly, scientific manner.

Within the robotics industry, each specialist contributes to the overall product. Generally a robot team consists of engineers, technologists/technicians, maintenance/service workers, sales engineers, and the technical educator or trainer. Each specialty requires a different training, education, and expertise level.

Robotics Engineer

Robotics engineers are usually background members of the robotics team. They research ways to design, build, and utilize robots for improved quality, productivity, cost savings, and safety. These engineers frequently contact management concerning cost proposals, efficiency studies, and quality control reports.

Robotics engineers conduct research and development activities concerned with the design, manufacture, and testing of electronic components, and robotic products and systems; furthermore, he or she develops applications of these products for commercial, industrial, medical, and scientific use. Robotics engineers design electrical circuits, electronic components, and integrated systems. In addition, they design and direct engineering personnel in the fabrication of test control apparatus and equipment, and they determine procedures for testing robotic products.

Robotics engineers must have at least a four-year college degree. Examples of management positions in engineering are the Manager of Robotics and the Director of Automation positions.

Robotics engineers should prefer activities dealing with things rather than people, and like scientific and technical activities and tasks carried on in relation to processes, machines, and techniques. Robotics engineers should be able to adapt to performing a variety of duties without loss of efficiency.

Robotic Technologist/Technician

Robot technologists/technicians (hereafter referred to as robot technologists) are in great demand in today's job market. Technologists take an engineer's ideas and plans to develop a workable robotic product. Technologists improve on engineers' ideas when necessary and serve as the link between the engineer and the development of the actual robotic product. While the engineer often works in the background, technologists actually implement the ideas.

Universities, colleges, technical institutions, and trade schools are all beginning to offer robot technology programs. Many colleges offer two- and four-year degree programs in robot technology. Technologist educational programs are basically less mathematically and theoretically based, and more hardware and process oriented, than engineering programs. Some of the courses required in a robotic technologist program are electronics, pneumatics, and hydraulics. The more years of training one has in robot technology, the higher salary one can expect to receive.

Robotic technologists should prefer activities dealing with things rather than people, and enjoy scientific and technical activities and tasks that are carried on in relation to processes, machines, and techniques. Robotic technologists should be able to perform a variety of duties, often changing from one task to another without losing efficiency, and to work in situations requiring precise tolerances or standards.

Robotic Maintenance/Servicer Person

Another occupation that has emerged within the new robotics industry is that of robot maintenance/servicer (hereafter referred to as one or the other, but not both). This job involves installing the robots and keeping them working. Maintenance professionals can be employed either by the robot distributor or manufacturer, or by the robot user. If servicers are employed by the robot manufacturer or distributor, they may be responsible for installation and service calls. Servicers employed in the robot manufacturing plant may service the initial production model of the robot and perform in-house maintenance and repair. These employees work closely with engineers and technicians.

Robotic servicers repair electronic equipment following blueprints and manufacturers' specifications, and align, adjust, and calibrate robotic equipment according to specifications. They test faulty robotic equipment and apply knowledge of the functional operation of electronic units and systems to diagnose the cause or causes of the malfunction. Robotic servicers also test electronic components and circuits to locate defects, replace defective components and wiring, and adjust mechanical parts.

The education required for the robot servicer is an apprentice training program and/or a degree from a two-year or four-year institution of higher learning.

Robotic service persons should enjoy activities involving business contact with people, activities concerned with the communication of data, and tasks resulting in tangible, productive satisfaction. Robotic maintenance persons should be able to perform a variety of duties, often moving from one task to another of a different nature in a short period of time. Work situations require precise tolerances or standards; thus, a robotic servicer should be able to cope with preciseness.

Robot Sales Representative

Many robotics companies offer positions for people trained in marketing and business administration. The robotics industry needs competent, informed salespeople. A good salesperson must be able to communicate effectively, understand people and their actions, and most importantly, know every detail of the product being represented. It is difficult to sell a robotic product if one cannot answer a buyer's technical questions. A sales representative must be able to give a detailed technical explanation of how the product works, determine what product is best suited for a particular plant's operation, and distinguish whether a robot would improve productivity and quality in a particular operation. To answer all these questions, the sales representative must be familiar with some aspects of robotics and engineering, as well as the work technologists and maintenance workers perform. The sales representative links the buyer to the robotics company, and only a qualified communicator and technically oriented person can fill this role. People with these qualities are in high demand.

Training in a two-year or four-year educational institution in marketing and business administration is recommended for the robotic sales representative. It is also suggested that the student take any basic robot courses that are offered.

The robotic sales representative should enjoy activities concerned with the communication of data, and scientific and technical activities. Robot sales representatives should be able to accept responsibility for the direction, control, and planning of an activity. To influence people in their opinions, attitudes, and judgments about ideas or things summarizes the core of a sales representative's job.

Robot Educator/Trainer

The robot educator/trainer (hereafter referred to as the robot educator) conducts training programs, confers with management, prepares a teaching outline, and selects and develops teaching aids. As the robotics industry expands and becomes more complex, the lack of people in the field to teach and train the personnel required by industry becomes more and more evident. Engineers are quickly developing more advanced, more complex robots. With the creation of these advanced machines comes the need for more qualified personnel to install and service them. Therefore, more educators are needed to train these service people.

The robot educator prepares and conducts training programs for the employees of industrial users of robots. The robot educator confers with management to gain knowledge of an identified work situation requiring preventive or remedial training for employees. The robot educator then formulates a teaching outline in conformance with selected instructional methods. He or she utilizes knowledge of specified training needs and the effectiveness of such training methods as individual coaching, group instruction, lectures, demonstrations, conferences, meetings, and workshops. The robot educator selects or develops such teaching aids as training handbooks, demonstration models, multimedia visual aids, and reference works. In addition, the robot educator tests trainees to measure their learning progress and to evaluate the effectiveness of training presentations.

In the past, because of the great demand for training people in the robotics industry, educators left universities and technical schools and accepted training positions within industry. Today, funds contributed by the government and the robotics industry allow schools to offer educators more attractive incentives to remain and meet the educational needs of people training for a career in the industry. Many robot companies are also establishing in-house training programs for new employees.

A variety of educators and trainers are needed at the plant level, in universities, colleges, trade and technical schools, vocational schools, and even in high schools to prepare workers for jobs in the robotics industry.

A two- or four-year course of study in robotics can lead to a variety of employment opportunities as a robotics educator.

Robot educators should excel in activities concerned with the communication of data and involving business contact with people. Robot educators should be able to accept responsibility for the direction, control, and planning of an activity, and to deal with people beyond the giving and receiving of instructions.

Summary

There are a variety of occupations associated with the robotics industry. They have been created as a result of this industry's emergence. There are also many diverse career opportunities for students of business administration, industrial engineering, electrical engineering, chemical engineering, mechanical engineering, and computer science. Employment opportunities will also exist for people who have attained skills from previous work experiences and are willing to be retrained to apply their skills in the robotics industry.

CHAPTER THREE

SPACE TECHNOLOGY

The Space Technology Industry

The space technology industry, unlike other high technology industries, is not new. Space technology experienced its major boom and growth period twenty-five years ago. What is new and different about this industry is that work has shifted from individual efforts to teams of specialists. Another significant change has been the development of space-based products.

The growth rate of the space workforce has also changed. Ten years ago there was a 15 percent annual growth rate of employees in, for example, TRW's space technology division. Today there is only a one to three percent annual growth rate. This sector of high technology is not currently expanding employment opportunities.

Twenty years ago it would have taken only three specialists to create an entire communication satellite. The work team was small and cohesive. Today it takes nearly 70 specialists organized in a diversified work team to develop one satellite. Each specialist works on a specific aspect of the satellite for which he or she is particularly well trained.

In the past, the space technology industry developed a smaller group of products. Unlike the multitude of products developed today in such companies as TRW, McDonnell Douglas, and Atlantic Research Corporation, each company only developed a few specialized products. Today, Atlantic Research Corporation not only develops and produces space technology products, but it also develops data communication products. Further, while McDonnell Douglas develops domestic and military aircraft and spacecraft, they explore new business opportunities in the area of space-based pharmaceuticals production.

The products in space technology industries have undergone a twofold evolution. First, twenty-five years of research has resulted in products that are less expensive, more reliable, of a higher quality, and more producible. Second, now that the industry has stabilized, individual corporations are refining techniques and finding new and different uses for existing products, rather than creating new products. The range of applications for specific products has also expanded. For example, the antenna was initially used only for television reception; now it is also used for cable television reception. Another example is the many products developed for and used in the Vietnam war that have now been adapted for domestic use. An infrared sensor developed to detect soldiers by their body heat is now used by fire departments and forest services to detect people in burning buildings and forests. Hip pocket lasers, developed to enable battleships to aim guns more accurately, are now used on construction sites for leveling, thus replacing the traditional transit. The communication satellite still appears much like the satellite of twenty years ago, but it too has evolved. Now more versatile and more manufacturable, it serves as a land/sea navigational aid, and facilitates telecommunications and earth resource mapping.

As in other industries, need often creates the product. It has been said that if the USSR had not placed the Sputnik satellite into space in the 1950s, the United States might have taken longer to develop a satellite and the ability to place a man on the moon, even though the technological capability existed. Currently some space technology companies are working towards developing alternative sources of energy and perfecting saltwater desalinization.

Required research and development for these projects, however, may not be adequately funded until the United States runs out of petroleum or major freshwater sources run dry. Because of this factor, space technology will continue to concentrate on the enhancement of, and alternative uses for, existing products.

Careers in Space Technology

The space technology industry offers many different areas for work specialization. Careers exist in administration, analytical work, design, development, research and testing, laboratories, computing, and production support. While still in high school, a student planning to enter the space technology industry should take courses in higher mathematics, physics, biology, and chemistry, along with other general education courses.

Although space technology companies use different job titles, the same occupations can generally be found in most companies.

Engineering-Development Technician

Engineering-development technicians apply electronic theory, principles of electrical circuits, electrical testing procedures, engineering mathematics, and physics to their work. They lay out, build, test, troubleshoot, repair, and modify such electronic equipment as computers, control instrumentation, and machine tool numerical controls that develop products in this industry. Engineering-development technicians discuss layout and assembly problems with electronics engineers and draw sketches to clarify design details and the functional criteria of electronic units. They assemble experimental circuitry or complete prototype models according to engineering instructions, technical manuals, and knowledge of electronic systems and components and their functions. They recommend changes in circuitry or installation specifications to simplify assembly and maintenance; they also set up standard test apparatus or contrive test equipment and circuitry to conduct functional, operational, environmental, and life tests that evaluate the performance and reliability of the prototype or production models. Analyzing and interpreting test data; adjusting, calibrating, aligning, and modifying circuitry and components; and recording effects on unit performance are other job duties of engineering-development technicians. They also write technical reports and develop charts, graphs, and schematics to describe and illustrate system operating characteristics, malfunctions, deviations from design specifications, and functional limitations for consideration by professional engineering personnel in broader determinations affecting systems design and laboratory procedures.

This position requires at least a four-year degree (B.S.) in engineering. The beginning salary ranges approximately from $21,000 to $24,000 per year.

The engineering-development technician should prefer activities dealing with things rather than people, and enjoy scientific and technical activities and tasks that are carried on in relation to processes, machines, and techniques. Engineering-development technicians should be able to make evaluations and recommendations based on test data.

Engineering-Document-Control Clerk/Configuration Management

Engineering-document-control clerks compile and maintain control records and related files in order to control the release of blueprints, drawings, and engineering documents to manufacturing and other operating departments in the space technology industry. They examine such documents as blueprints, drawings, and specifications of space technology products and services to verify the completeness and accuracy of data, and confer with document originators or engineering liaison personnel to resolve discrepancies. They compile the required changes to these documents and then post the changes to control records, release documents, and notify affected departments.

Engineering-document-control clerks should enjoy activities concerned with the communication of data, and activities of a routine, concrete, organized nature. Engineering-document-control clerks should be able to make generalizations, evaluations, and decisions based on measurable or verifiable criteria, and work in situations requiring precise tolerances and standards.

Guidance Navigation and Control Engineer

Guidance navigation and control (GNC) engineers provide technical support for the GNC system design, and development and analysis for satellites and space launch vehicles. Such engineers develop basic design concepts used in the design, development, and validation of electromechanical GNC systems for stabilizing, navigating, and maneuvering vehicles in flight.

Guidance navigation and control engineers should have capabilities and experience in the following areas: rigid body dynamics, guidance, digital flight controls, control law development and simulation, flight control computer software validation, stability and controls analysis, estimation, and filtering. An M.S. or Ph.D. degree in electrical/electronic engineering or aeronautical engineering is required. The beginning salary ranges between approximately $34,000 and $42,000 per year.

Guidance navigation and control engineers should enjoy scientific and technical activities, and tasks of an abstract and creative nature. Being able to change from one task to another of a different nature without losing efficiency, and accepting responsibility for the direction, control, or planning of an activity, is necessary.

Guidance and Control Electronics/Microprocessor Engineer

Guidance and control electronics/microprocessor engineers (hereafter referred to as guidance and control electronics engineers) analyze and design analog and digital circuits, as well as microprocessor applications, for current and future launch vehicle and spacecraft guidance and control systems. They test models, prototype subassemblies, or production vehicles to evaluate the operational characteristics and effects of environmental stress imposed during actual or simulated flight conditions. Guidance and control electronics engineers may also specialize in analytical programs aimed at parametrically defining the operational characteristics of various electronics subsystems.

A working knowledge of computer-aided circuit analysis programs, guidance, control system interfaces, and microprocessor hardware and software, plus an M.S. or Ph.D. degree in electrical/electronic engineering or aeronautical engineering is required of guidance and control electronics engineers. The beginning salary ranges from approximately $34,000 to $42,000 per year.

Guidance and control electronics engineers should enjoy technical analysis and testing activities, and tasks of an abstract and creative nature. Performing a variety of duties, often changing from one task to another without losing efficiency, is characteristic of this field. Accepting responsibility for the direction, control, and planning of an activity at the project level is an early requirement for candidates in this area. The guidance and control electronics engineer should also be able to make generalizations, evaluations, and decisions based on derived analytical or acquired test data.

Satellite Systems Engineer

Satellite systems engineers provide technical support for new and existing satellite programs by designing, developing, and integrating new systems, and applying engineering principles at both project and functional levels.

An M.S. or Ph.D. degree is required in electrical/electronic or aeronautical engineering. The beginning salary ranges from approximately $27,000 to $38,000.

Satellite systems engineers should possess strong scientific and technical capabilities, and should enjoy interacting with a variety of technical organizations. Satellite systems engineers should be able to perform a variety of duties, often changing from one task to another of a different nature without losing efficiency, and accept responsibility for the direction, control, and planning of an activity.

Control System Analyst

Control system analysts analyze the performance of satellite-related control systems. They evaluate models, prototypes, subassemblies, and production space vehicles to insure that all requirements for controllability and maneuverability are satisfied during actual or simulated flight conditions. Control system analysts also specialize in designing and developing computer analysis methods for analytical evaluations of control systems.

The areas of technical interest for control systems analysis include the design and modeling of spacecraft attitude control and measurement systems, the formulation of dynamic equations describing multibody mechanical systems, and the analysis and simulation of control systems. An M.S. or Ph.D. degree in electrical/electronic engineering or aeronautical engineering is required.

Control systems analysts should take an interest in scientific and technical activities, and like tasks of an abstract, creative nature. Being able to perform a variety of duties, changing from one task to another of a different nature without losing efficiency, and accepting responsibility for the direction, control, and planning of an activity are important abilities for control system analysts.

Real-Time Simulation Programmer

Real-time simulation programmers develop real-time digital simulations primarily using the FORTRAN programming language with a minor amount of Assembly language code. Programmed applications simulate dynamic spacecraft systems and involve interfaces with hardware and human controllers.

Real-time simulation programmers must have an M.S. or Ph.D. degree in electrical/electronic engineering or aeronautical engineering.

Real-time simulation programmers should enjoy scientific and technical activities, and like tasks of an abstract, creative nature. Performing a variety of duties, often changing from one task to another of a different nature without loss of efficiency, and accepting responsibility for the direction, control, and planning of an activity are necessary abilities for real-time simulation programmers.

Performance Analysis Engineer

Performance analysis engineers perform propulsion systems and vehicle flight performance analyses relating to launch and re-entry vehicles. They test models, prototypes, subassemblies, and the production of space vehicles to study and evaluate operational characteristics and the effects of stress imposed during actual or simulated flight conditions. Performance analysis engineers may specialize in analytical programs concerned with ground or flight testing, or the development of acoustic, thermodynamic, or propulsion systems. Capabilities and experience in launch vehicle modeling and simulation, re-entry dynamics, and flight performance optimization are also necessary.

An M.S. or Ph.D. degree in electrical/electronic engineering or aeronautical engineering is required. The beginning salary ranges approximately from $25,000 to $36,000 per year.

Performance analysis engineers should enjoy scientific and technical activities, and tasks of an abstract, creative nature.

Satellite Attitude Estimation and Control Analyst

These analysts apply modern estimation and control system synthesis methods to satellite attitude and position determination. Using optimal control theory and stochastic estimation techniques, these analysts work closely with control system analysts to ensure that requirements for precise attitude and position control are met. They perform analysis and design work on advanced high performance pointing and tracking servo-mechanisms, including the application of laser systems.

An M.S. or Ph.D. degree in either electrical/electronic engineering or aeronautical engineering is required. The salary ranges approximately from $26,900 to $46,900.

Satellite attitude estimation and control analysts should enjoy scientific and technical activities, and tasks of an abstract, creative nature. For these analysts, performing a variety of duties, changing from one task to another of a different nature without losing efficiency, and accepting responsibility for the direction, control, and planning of an activity is important.

Laser Communication Engineer

A background in communications systems, and a knowledge of the various laser transmitter technologies (including solid state and gas) as well as laser receiver and optical components are required to be a laser communication engineer. An M.S. or Ph.D. degree in engineering or physics is also necessary. The beginning salary ranges from approximately $18,500 to $28,500.

Laser communication engineers should enjoy scientific and technical activities, and tasks that are carried on in relation to processes, machines, and techniques. Performing a variety of duties, often changing from one task to another of a different nature without losing efficiency, is also necessary to be a successful laser communication engineer.

Optical and Electro-Optical Systems Analysis Engineer

Optical and electro-optical systems analysis engineers perform analysis and laboratory evaluations of optical and electro-optical components and systems used within space vehicles. Signal processing devices, lasers, laser modulators, fiber and integrated optics, coherent optical processing, image analysis, and diffraction effects are also optical systems that optical and electro-optical systems analysis engineers design.

Optical and electro-optical systems analysis engineers analyze optical systems to fit within the specified physical limits of precision optical instruments used in space vehicles. They determine the specifications for operations and make adjustments to calibrate and obtain specified operational performances. Conducting research on light-emitting and light-sensitive devices, and designing electronic circuitry and optical components with specific characteristics to fit within specified mechanical limits and perform according to specifications are also responsibilities of optical and electro-optical systems analysis engineers.

An M.S. or Ph.D. degree in engineering or physics is required. Beginning salaries range from approximately $24,000 to $32,000 per year.

Optical and electro-optical systems analysis engineers should like scientific and technical activities, and tasks that are carried on in relation to processes, machines, and techniques. Accepting responsibility for the direction, control, and planning of an activity, and making generalizations, evaluations, and decisions based on measurable or verifiable criteria are important abilities for these engineers. Optical and electro-optical systems analysis engineers should be able to adapt to work situations requiring precise tolerances or standards, and to performing a variety of duties, often changing from one task to another of a different nature without loss of efficiency.

Engineering Systems Analyst

Engineering analysts analyze and evaluate state-of-the-art satellite systems to develop advanced concepts, and design next-generation systems.

These analysts conduct logical analyses of scientific, engineering, and other technical space problems. Consulting with engineering or scientific personnel enables the engineering systems analyst to refine the definition of the project and prepare a mathematical simulation of the space system under study.

Analysts must have a solid academic background and/or industrial experience in at least two of the following disciplines: modern control theory, dynamic programming, estimation theory and applications, stochastic processes, astrodynamics communications theory, signal analysis, electro-optics, and radar systems. An M.S. or Ph.D. degree in mathematics or computer science is required.

Engineering systems analysts should be interested in scientific and technical activities and tasks that are carried on in relation to processes, machines, and techniques. Adapting to performing a variety of duties, changing from one task to another of a different nature without losing efficiency, and making generalizations, evaluations, and decisions based on sensory or judgmental criteria, or on measurable or verifiable criteria is required.

Image Processing Analyst

Image processing analysts are involved in the analysis, evaluation, and specification of image processing systems in space technology, encompassing the imaging sensor, sensor signal processing and digitization, digital communications, and final image manipulation, exploitation, and interpretation. This work encompasses algorithm research and simulation in the areas of image restoration and enhancement, coding, quantization detection and estimation, pattern recognition and image understanding, and multispectral image correlation and exploitation.

An image processing analyst should have a working knowledge of digital and analog communications theory, probability theory, Fourier optics, and FORTRAN programming. An M.S. or Ph.D. degree in mathematics or computer science is also required.

Scientific Programmer/Analyst

Scientific programmers/analysts (henceforth referred to as scientific programmers) must have knowledge of and experience with the FORTRAN programming language, as well as experience with large, high-speed digital computers, including structured program design and analysis.

Scientific programmers convert scientific, engineering, and other technical problem formulations concerning space technology research and development to formats processible by computers. Scientific programmers resolve symbolic formulations, prepare flow charts and block diagrams, and encode resultant equations for computer processing. They apply knowledge of advanced mathematics and an understanding of computer capabilities and limitations to problems in the space technology industries. Scientific programmers then review the results of computer output with other project team members to determine the necessity for modifications and reprocessing.

A background in one or more of the following disciplines is required: mathematical analysis, numerical analysis, computer architecture, modern operating systems, or direct experience with at least one engineering or physical science application area. Scientific programmers must have an M.S. or Ph.D. degree in mathematics or computer science. Beginning salaries fall between $24,500 and $37,500.

Scientific programmers should enjoy scientific and technical activities, and tasks of an abstract, creative nature. Making generalizations, evaluations, and decisions based on sensory or judgmental criteria, or on measurable or verifiable criteria is an important responsibility of scientific programmers.

Software Engineer

Software engineers analyze space problems, plan and develop software programs, transfer programs to memory chips, install chips on printed circuit boards, and test and correct the operation of chips and boards in on-board computers, using computer equipment.

They must have a strong interest in developing state-of-the-art software for space systems, in investigating methods to bridge the gap between software engineering theories and practices, and in evaluating the applications of new advances in software engineering and computer technology as they relate to space technology. This position requires an understanding of hardware functions to ensure that space system requirements for software can be accommodated. Software engineers are involved in the definition, evaluation, and acquisition of computing resources embedded in or associated with advanced space systems.

Software engineers need a working knowledge of computer hardware—from microprocessors to large-scale systems—and of Assembly and higher-order languages. The software engineer is required to have an M.S. or Ph.D. degree in mathematics or computer science.

Other Career Opportunities

As can be seen from the occupations described in this chapter, there are many fields of work specialization within the space technology industry. A new employee with the necessary qualifications in one specialty area will be assigned to a small project or program team. The team will expect the new employee to blend experience and education with that of the other team members.

If a person's major area of study is chemical engineering or chemistry, positions can be found within the space technology industry in technical writing and illustrating, aircraft engineering, instrumentation, materials and processes, support equipment, and quality assurance. If a person's area of study is mathematics or statistics, then positions may be found in engineering administration, control systems, flight test operations, tracking and control, simulation systems, and computer programming. The person with a major area of study in business administration, economics/law, or marketing may find a position in the areas of accounting, electrical power systems, operations research/analysis, computer applications designs, facilities engineering, and value engineering. This list is far from complete and there are many other areas of study that can be applied to occupations within the space technology industry.

The space technology industry, while not a new emerging industry, has evolved and its workforce has changed with the industry. Even though the industry may not be growing as fast as it was twenty-five years ago, it still needs educated, trained, and competent people.

CHAPTER FOUR

TELECOMMUNICATIONS

The Telecommunications Industry

WITH THE EMERGENCE OF THE INFORMATION AGE, THE TELECOMMUnications industry has expanded rapidly. The information age succeeds the industrial age, which succeeded the agricultural age. The industrial age surpassed the agricultural age in the early 1900s, when the number of people engaged in industry began to exceed those in the production of food. In the last 25 years, the number of people employed in jobs associated with the processing of information has exceeded those employed in industry. The information industry is now the major employer in developed nations. More and more people are working for the information industries, which include education, publishing, accounting, journalism, printing, television, radio, banking, and telecommunications. As telecommunications technology becomes increasingly linked to computers, the information base, and thus the information industries, will continue to expand.

Telecommunications companies perform research and development, manufacture, sell, and install and maintain a wide range of products. They also offer a variety of services. Their primary products and services include local telephone service, Yellow Pages advertising, cellular mobile radio telephone service, communications equipment, systems and support services (including digital switching and PABX systems), transmission equipment, residential and business telephone products, long distance telephone services, nationwide data communication networks, photographic lighting products, lamp fixtures, and related products.

Today a new telecommunications industry is evolving and expanding through changes caused by the rapid development of new communications and computer technologies. Additional factors influencing the industry are the emergence of promising new markets for products and services employing these new technologies, increased competition in the telecommunications market stimulated by decisions from government and the courts, and deregulation of key elements in the nation's telephone business that began in January 1983.

Since the first days of telecommunications, ground-based technology (called terrestrial microwave) has been used for communication systems. This ground-based technology has given way recently to the use of space-based satellite systems. Satellites bounce signals all over the world much more rapidly and less expensively than the old ground-based method of stringing wire from point A to point B.

Communication companies, like RCA, have developed new hybrid satellites that they started sending into space in late 1985. Hybrid satellites use both the old C-Band technology, and the new Ku-Band technology. *Ku-Band,* sometimes called direct broadcast, carries a high-quality signal to small antennas, six feet or less in diameter. One example of a commercial use is the condominium or apartment complex that uses one satellite dish to receive signals from hybrid satellites. All the tenants can be linked to the satellite dish by copper wire and receive cable television channels from around the world.

Satellite systems are less expensive than ground-based/terrestrial microwave systems. They have the ability to receive and send high quality signals anywhere, thus making substantially lower rates possible for coast-to-coast private line service, both voice and data, and stereo audio/video transmission.

Beyond providing less expensive rates, satellite technology has helped expand the communications industry. RCA is helping other companies expand by selling them satellite transmission capabilities. MCI, for example, leases many channels from RCA for their domestic resale operations and is consequently able to compete successfully with larger companies like AT&T. Using this satellite technology MCI has grown to a large company employing approximately 350,000 people. Thirty percent of MCI personnel work in technical areas, thirty percent in sales support, and the remainder in direct sales. Cable television is another example of growth made possible by the advent of satellite technology. Satellites enable cable television companies to address an unlimited number of receiving stations around the country simultaneously. Videotex is another technology expanding the telecommunications industry. Videotex technology enables specially adapted television sets to access a computer data bank and call up textual or graphic material onto the television screen. Videotex can be used by consumers to look for a job, check weather reports, or transfer funds from one bank account to another without leaving home.

Careers in Telecommunications

The two major career paths in the telecommunications industry are the sales and technical areas. A sales team consists of account executives, technical consultants, and administrative support personnel. Technically based occupations are filled by chemists, materials engineers, physicists, electro-optics engineers, electrical engineers, electronics engineers, computer scientists, mechanical engineers, and industrial engineers.

Account Executive

An account executive sells communication equipment and services to the public and private business sectors, and is the primary sales contact on a group of assigned accounts representing a territory.

The account executive is measured on quota achievement and is accountable for maintaining territory control through sales account planning and forecasting, territory planning, performance and administration, and customer satisfaction. The account executive directs the efforts of technical and administrative resources in the sales implementation process. The account executive usually works on a commission-based compensation plan.

An account executive must complete a company offered certification process within a specified period of time that requires demonstrated performance in such areas as basic industry knowledge, product knowledge, selling skills, resource utilization, and successful sales efforts.

The account executive should enjoy activities concerned with the communication of data and activities involving business contact with people.

Technical Consultant

Technical consultants contact residential, commercial, and industrial telephone company subscribers to solve communication problems and needs, and to promote the use of telephone services by utilizing knowledge of marketing conditions, contracts, sales methods, and communications services and equipment.

Technical consultants are responsible for analyzing customers' information and communications systems elements; developing specific technical designs and recommended system configurations involving voice, data, energy, and network arrangements; and assisting the sales team in preparing and presenting specific sales proposals. Technical consultants work closely with customers to insure that systems sold by the account executive are appropriately designed. Working with the supervisor (systems manager), technical consultants must demonstrate skills as project managers in order to coordinate the efforts of other team members and departments to assure the completion of key activities, from design through the implementation phase of system sales. Technical consultants also consult with customers and other team members regarding ongoing system expansions or reconfigurations and hardware changes.

If a technical consultant demonstrates technical abilities, the career path can eventually lead to systems management and the supervision of groups of technical consultants, or to a more advanced level of consulting that involves the highest level of complexity in analysis and design.

A technical consultant must have at least a B.A. or B.S. in business administration. Technical consultants should gravitate toward activities concerned with the communication of data and activities involving business contact with people. Technical consultants should be able to influence people in their opinions, attitudes, and judgments about ideas or things.

Chemist and Chemical Engineer

Most chemistry within the telecommunications industry is inorganic, but a wide spectrum of sub-disciplines are encompassed, including solid state chemistry, liquid extraction processes, plasma chemistry, laser-induced chemistry, analytical chemistry, and polymers. Chemists and chemical engineers design, research, analyze, and refine telecommunications products.

Chemists and chemical engineers design telecommunications equipment and develop processes for manufacturing chemicals and related products utilizing the principles and technology of chemistry, physics, mathematics, engineering, and related physical and natural sciences. They conduct research to develop new and improved telecommunications chemical manufacturing processes. By analyzing operating procedures and equipment and machinery functions, chemists and chemical engineers are able to reduce processing time and costs. Finally, they perform tests and take measurements throughout stages of production to determine the degree of control over such variables as temperature, density, specific gravity, and pressure.

Chemists and chemical engineers must have at least a B.S. degree in chemistry or chemical engineering. Starting salaries range from $20,000 to $29,000 per year.

Chemists and chemical engineers should be interested in activities of a scientific and technical nature, and activities that are carried on in relation to processes, machines, and techniques. Chemists and chemical engineers should be able to adapt to performing a variety of duties, often changing from one task to another of a different nature without losing efficiency.

Materials Engineer

Materials engineering is interdisciplinary, incorporating metallurgy and ceramic science technology as well as chemistry and solid state science. Its purpose is to develop telecommunications processes and materials for specific industrial applications.

Materials engineers are involved in activities that include the development of alloys, intermetallic compounds, glass, organic polymers, semiconductors, and luminescent materials. The application and development of processing and characterization methods also falls to materials engineers. They evaluate and review plans, confer with producers of telecommunications materials, review data, interpret lab tests, and recommend actions.

Materials engineers evaluate technical and economic factors, recommending engineering and manufacturing actions for attaining the design objectives of processes or products by applying knowledge of material science and related technologies. They review plans for new telecommunications products by analyzing such factors as strength, weight, and cost, thereby insuring the attainment of design objectives. They plan and implement laboratory operations to develop material and fabrication procedures for new materials in order to fulfill product cost and performance standards. Materials engineers confer with producers of such materials as metals, ceramics, and polymers during the investigation and evaluation of materials suitable for specific telecommunications product applications; review product failure data; and interpret laboratory tests and analysis to establish or rule out material and process causes.

Materials engineers must have at least a B.S. degree in materials engineering. Starting salaries range from $23,000 to $26,000.

Materials engineers should find activities of a scientific and technical nature agreeable, and like activities that are carried on in relation to processes, machines, and techniques. Materials engineers should be able to perform a variety of duties, often changing from one task to another of a different nature without losing efficiency.

Physicist

Physicists are usually involved with assignments related to optics, solid state physics, or gaseous electronics. Physicists conduct research in phases of physical phenomena, develop theories and laws on the basis of observation and experiments, and devise methods in order to apply laws and theories of physics to telecommunications research.

Physicists perform experiments with masers, lasers, cyclotrons, telescopes, mass spectrometers, electron microscopes, and other equipment to observe the structure and properties of matter, the transformation and propagation of energy, relationships between matter and energy, and other physical phenomena. They then describe and express observations and conclusions in mathematical terms. Physicists also devise procedures for the physical testing of materials used in telecommunications products and conduct instrumental analyses to determine the physical properties of those materials.

Physicists must have at least a B.S. degree in physics. The starting salary is $25,000 per year.

Physicists should enjoy delving into activities of a scientific and technical nature. Physicists will find work situations requiring precise tolerances or standards to be the norm.

Electro-Optics Engineer

Electro-optics engineers conduct research, design work, and confer with engineering personnel regarding the use of electro-optics technology in telecommunications products.

Optical communications involves a broad spectrum of engineering skills covering classical optics, materials science, and mechanical design for the development of active and passive micro-optic components. The design and development of these components is performed within the context of systems integration into complete fiber optic data transmission links.

Electro-optical engineers conduct research, and plan the development and design of gas and solid state lasers, masers, infrared, and other light emitting and light sensitive devices that are used in new telecommunications products. Designing electronic circuitry and optical components with specific characteristics to fit within specified mechanical limits and perform according to specifications is a responsibility of electro-optical engineers. They also design, for telecommunications products, suitable mounts for optics and power supply systems, incorporate methods for maintenance and repair of components, and design and develop test instrumentation and test procedures. The electro-optical engineer confers with engineering and technical personnel regarding the fabrication and testing of prototype telecommunications systems, and modifies the design as required.

Electro-optics engineers must have at least a B.S. degree. The electro-optics engineer should be interested in activities of a scientific and technical nature, and activities that are carried on in relation to processes, machines, and techniques. Electro-optics engineers should be prepared to accept responsibility for the direction, control, and planning of an activity.

Electrical Engineer

Electrical engineers design equipment for all aspects of telecommunications systems. There is a heavy reliance on the use of microcompressors and other LSI (Large Scale

Integration) devices in the digital portion of new telecommunications equipment.

Electrical engineers conduct research and development activities concerned with the design, manufacture, and testing of the electrical components of telecommunications equipment and systems. They investigate how to apply this equipment to new uses as well as being involved with the manufacture, construction, and installation of electrically based telecommunications equipment, facilities, and systems. Electrical engineers design and direct the activities of engineering personnel engaged in the fabrication of test control apparatus and equipment; and in the establishment of methods, procedures, and conditions for testing telecommunications equipment. Electrical engineers are responsible for developing applications for controls, instruments, and systems to new commercial, domestic, and industrial uses of telecommunications. They also direct activities concerned with manufacturing, construction, installation, and operational testing to insure conformance of telecommunications equipment and systems with functional specifications and customer requirements.

Electrical engineers must have at least a B.S. degree in electrical engineering. Starting salaries range from $19,800 to $29,400 per year.

Electrical engineers should prefer activities dealing with things rather than people, and enjoy scientific and technical activities and tasks carried on in relation to processes, machines, and techniques.

Electronic Engineer

Electronics engineers conduct research on electronic phenomena relating to telecommunications systems.

This discipline interacts with electrical engineering activities relative to solid state circuit design and microelectronics with a preponderance of LSI (Large Scale Integration) and VLSI (Very Large Scale Integration) circuit work being performed for communications-related products. There is also a crossover in the area of computer science in that many of the tools for microelectronics design are developed in-house by computer science groups working in conjuntion with the microelectronics group. This situation also holds true for computer assisted drafting/computer assisted manufacturing (industrial engineering) personnel employed in large numbers for work in the software/microelectronics area.

An electronic engineer must have a B.S. degree. Engineers in this area should enjoy activities concerned with the communication of data, and activities of a scientific and technical nature.

Computer Scientist

Computer scientists formulate mathematical models of telecommunications systems, and set up and control analog or hybrid computer systems to solve scientific and engineering problems. They also observe the behavior of variables. After preparing mathematical models of telecommunications product and service problems, computer scientists draw data-flow charts to indicate the mathematical steps required in solving problems. In addition, computer scientists consult with the originator of a problem to determine the sources and methods of data collection and the methods of determining the values of variables. They may also examine and study physical models, graphic representations, and verbal descriptions of problems to apply knowledge of scientific disciplines and define problems. They draw computer-circuit diagrams to indicate connections between components and their values, and wire patchboard onto computers for the development of telecommunications services and products. Computer scientists observe behavior of variables on output devices, such as plotters, recorders, digital voltmeters, ocsilloscopes, digital displays, and readouts to obtain solutions to problems. They also prepare technical reports for the problem originator, describing the step-by-step solution to the given problem.

Computer scientists must have at least a B.S. degree in computer science.

Computer scientists should find scientific and technical activities interesting, and like tasks of an abstract and creative nature. Computer scientists should be able to accept responsibility for the direction, control, and planning of an activity; make generalizations, evaluations, and decisions based on measurable or verifiable criteria; and perform a variety of duties, often changing from one task to another of a different nature without losing efficiency.

Mechanical Engineer

Mechanical engineers in the telecommunications industry may be involved in activities ranging from product design and development to the design and analysis of high speed machinery for automatically assembling high-volume telecommunications products. The product range is broad and includes such diverse items as telephone components and equipment, electrical distribution products, microcircuits, structures, lighting products, high performance components made of precision materials, and equipment for satellite communications. The needed expertise is not limited to design alone, but includes mathematical analysis, manufacturing and processing, electromechanics, thermodynamics, systems applications, and testing.

Mechanical engineers plan and design mechanical or electromechanical telecommunications products or systems, and direct and coordinate the operation and repair activities in the telecommunications industry. Mechanical engineers design telecommunications products or systems utilizing and applying knowledge of engineering principles. They also plan and direct engineering personnel in the fabrication of test control apparatus and equipment, and in the development of methods and procedures for testing telecommunications products or systems. Furthermore, the mechanical engineer directs and coordinates the construction of telecommunications systems and products and installation activities. Insuring that products and systems conform with engineering design and customer specifications is another job responsibility of mechanical engineers.

Mechanical engineers must have at least a B.S. degree in mechanical engineering. The starting salary range falls between $20,000 and $29,000.

Mechanical engineers should prefer activities dealing with things rather than people, because activities of a scientific and technical nature and activities carried on in relation to processes, machines, and techniques comprise the bulk of the work. Mechanical engineers should be able to perform a variety of duties, often changing from one task to another of a different nature without losing efficiency. Responsibility for the direction, control, and planning of an activity often falls to the mechanical engineers.

Industrial Engineer

The industrial engineer's job has been defined as the accomplishment of tasks in the least possible time, at the lowest total cost, with the most satisfaction going to the people performing the work. It involves the application of modern techniques to the design and processing of telecommunications materials to bring new manufacturing approaches into reality. Industrial engineers plan the utilization of production facilities and personnel to improve the efficiency of operations in a telecommunication industrial establishment.

Industrial engineers study functional statements, organization charts, and project information to determine the functions and responsibilities of various workers and work units in a telecommunications company. Industrial engineers identify areas of overlap in duties and responsibilities, establish work measurement programs, and make sample observations of work to develop standards of manpower utilization in the telecommunications industries. Industrial engineers also analyze the utilization of manpower and machines in units, and develop work simplification programs in such areas as work distribution, work count, process flow, economy of worker motions, and layout of units in telecommunications companies. Finally, industrial engineers plan space layout of units to ascertain the objectives of work measurement and simplification studies.

Industrial engineers must have at least a B.S. degree in industrial engineering. Starting salaries range from $18,700 to $27,900.

Industrial engineers should enjoy scientific and technical activities, and tasks that are carried on in relation to processes, machines, and techniques. Industrial engineers must be able to perform a variety of duties, often changing from one task to another of a different nature without losing efficiency; and accept responsibility for the direction, control, and planning of an activity.

Summary

The telecommunications industry has received a definite boost from high technology. Many diverse occupations exist in the telecommunications industry. Because there are so many more products currently available to sell and customers to sell to, the employment opportunities especially within the sales area are increasing more than those within the technician areas.

As satellites and other new telecommunications systems become more sophisticated, and are used for more and more consumer services, the telecommunications industry will continue to expand. The use of the computer and advancing high technology create less expensive and easier to use systems, such as those used for home banking. They also modernize such systems as the telephone services. Without question the telecommunications industry is a strong industry that will continue to grow.

CHAPTER FIVE

BIOTECHNOLOGY

The Biotechnology Industry

THE MEDICAL FIELD, WHICH INCLUDES THE BIOTECHNOLOGY industry, has never stopped growing. With each new discovery, from penicillin to recombinant DNA technology, the medical field has expanded. As biotechnology continues to expand with new products and services, there will be a need for more personnel trained as physicists, chemists, genetic engineers, lab technicians, nurses, and other health professionals.

Modern medicine began in the second half of the 19th century when Louis Pasteur proved that infectious diseases were caused by microscopic living organisms invading the body. The work of Pasteur and others in the science of bacteriology was followed in the 20th century by scientifically based drug therapy such as the use of penicillin.

We are entering another era in the 1980s, the era of molecular medicine. Today, through the use of recombinant DNA technology and other scientific advances, the potential exists to harness the body's own mechanisms for treating diseases and disorders with new, more therapeutic substances. Recombinant DNA technology enables a specific human gene to be isolated, then spliced into the DNA of a microorganism. This technique is accomplished in such a way that the organism follows the gene's instructions for making a desired protein. When the organism reproduces, it passes on these instructions to its offspring, which also make the protein. Biotechnology firms can now mass produce proteins that were difficult or impossible to produce by more conventional techniques in the past. The advantages to human and animal health because of this new technology are as follows:

- Recombinant DNA technology creates an invaluable research tool for better understanding biological mechanisms.

- Recombinant DNA technology makes possible the development of new and better ways of making therapeutics that are already in use.

- Recombinant DNA technology can provide therapeutic products that were previously in short supply and completely new categories of drugs not available in the past.

There are endless possibilities for the use of recombinant DNA technology. Four examples of present uses are presented so the reader can understand the significance of this new technology.

Recombinant DNA technology is being tested for use on a recombinant human growth hormone for treating short stature due to this hormone deficiency.

Gamma interferon is being tested as an anti-cancer drug. Preclinical studies indicate the product may directly inhibit the growth of cancer cells and may be more effective than other types of interferon in stimulating the body's immune system to fight cancer. Gamma interferon may also be used in the treatment of viral and parasitic diseases.

Tissue-type Plasminogen Activator (t-PA) is a protein normally produced by the body to aid in the dissolution of blood clots. Recombinant t-PA is being tested for use in dissolving blood clots in people suffering from heart attacks and other life-threatening conditions. According to preclinical data, the product may be safer and more effective than current clot dissolving agents because it appears to be clot-specific.

Bovine interferons are naturally occurring proteins from the immune system in cattle. The testing of the recombinant product indicates that bovine interferons offer potential protection against many viral-related diseases that result in hundreds of millions of dollars in losses to U.S. ranchers each year.

Thus, like the discovery of the relationship between microbes and disease and the subsequent introduction of antibiotics and synthetic drugs, molecular medicine (recombinant DNA) represents a large expansion and change in medical science. This biotechnology of the 1980s may possibly change forever the way in which diseases and disorders are treated.

Some of the new biotechnology firms are making test kits that physicians can use in their offices or laboratories and, in some cases, consumers can use at home. These test kits are easy to use and eliminate the need for elaborate laboratory instrumentation and expensive laboratory time. An example is the pregnancy test kits developed by Monoclonal Antibodies, Inc. in California. These test kits increase sensitivity and eliminate interfering substances. The kit produces a plainly visible color change in a test tube or on a white dipstick if pregnancy is indicated. The results are obtained more quickly, taking only twenty minutes to one and a half hours.

Another product that has been developed by the biotechnology industry is a kit that tests and predicts the timing of human ovulation. This test measures the rise in levels of a hormone found in urine that triggers the release of the egg from the ovary approximately 24 to 36 hours later. While tests were available in the past to measure this hormone, they employed sophisticated radio immunoassay techniques requiring radio isotopes and expensive equipment. Therefore, they could only be performed in well-equipped laboratories from which response time was typically 24 hours or more. Such a delay in test results generally rendered this information of little practical value in family planning. This new test, in contrast, permits measurements to be made in 80 minutes, without the need for any instrumentation or equipment, in a physician's office. This product is designed to provide a simpler, more reliable, and more effective prediction of ovulation as a fertility aid.

Many biotechnology companies have devoted a great deal of their resources to research and development in recombinant DNA technology, and in the manufacturing and marketing of their resulting products. As the demand for these new products and services grows, biotechnology companies are doubling and tripling their work force. As a result, this new technology is creating new jobs and opening up the need for more personnel in already existing jobs.

Careers in Biotechnology

The next section will describe the new and in-demand occupations created by the biotechnology industry. The occupations will be described. The required education and training will be explained, and the abilities needed by workers in the industry will be outlined. Information on wages paid in the biotechnology industry is not available.

Associate Scientist/Scientist

Associate scientist/scientists (henceforth referred to as associate scientists) study chemical processes, conduct research, and work with protein isolation and characterization in the research and development of biotechnology products and services.

Associate scientists study the chemical processes of living organisms, examine the chemical aspects in the formation of antibodies, and conduct research into the chemistry of cells and blood corpuscles. Associate scientists also study the chemistry of living processes, including the mechanisms of development of normal and abnormal cells, breathing and digestion, and of such living energy changes as growth, aging, and death to understand more fully what products and services need to be developed. In addition they conduct research to determine the action of foods, drugs, serums, hormones, and other substances on tissues and vital processes of living organisms. Finally, associate scientists isolate, analyze, and identify hormones, vitamins, allergens, minerals, and enzymes, and determine their effects on body functions.

Associate scientists are required to have a Ph.D. in biochemistry or chemistry or an M.D. degree, and two or more years experience in protein chemistry.

Associate scientists should be interested in scientific and technical activities. They must be able to make generalizations, evaluations, and decisions based on measurable or verifiable criteria. Work situations requiring precise tolerances or standards must be tolerated.

Senior Biostatistician

The senior biostatistician directs the biostatistical and data handling functions of clinical and pharmacology groups. This job is done by planning data collection, and analyzing and interpreting numerical data from experiments, studies, surveys, and other sources. He or she also applies statistical methodology to provide information for scientific research and statistical analysis. This information is then used in the manufacture and development of biotechnology products.

The senior biostatistician plans methods to collect information and develops questionnaire techniques for the biotechnology company according to survey design. He or she then conducts and analyzes surveys utilizing sampling techniques or complete enumeration bases. The senior biostatistician also evaluates and interprets the reliability of source information, adjusts and weighs raw data, and organizes results into a form compatible with analysis by computers and other methods. Information is presented by the senior biostatistician through computer readouts, graphs, charts, tables, written reports, and other methods.

The senior biostatistician is required to have a Ph.D. in biostatistics and at least three years postdoctoral experience in setting up and running clinical studies and interacting with the Food and Drug Administration.

The senior biostatistician should be interested in scientific and technical activities and be able to make generalizations, evaluations, and decisions.

Scientific Programmer

Scientific programmers provide general scientific programming support to the research and development area in the biotechnology company. They also interact with scientists; define solutions to problems; and code, test, debug, and install software changes required to implement solutions.

Scientific programmers convert scientific, engineering, and other technical problem formulations to a format processable by computer; resolve symbolic formulations; prepare flow charts and block diagrams; and encode resulting equations for processing—all for use in the biotechnology company. They apply knowledge of such advanced mathematics as differential equations and numerical analysis, with an understanding of computer capabilities and limitations.

Scientific programmers must have a B.S. degree and the ability to work on many small projects with minimum supervision.

The scientific programmer should enjoy scientific and technical activities, and tasks of an abstract and creative nature.

Laboratory Associate/Research Assistant

Laboratory associates/research assistants (henceforth referred to as laboratory associates) conduct chemical and physical laboratory tests in the biotechnology company, and make qualitative and quantitative analyses of materials, liquids, and gases for research, development of new products and materials, processing and production methods, quality control, and maintenance of health and safety standards. They conduct general biochemical experiments, and assist in the development of therapeutic protein products. Laboratory associates, under general supervision, perform a wide variety of laboratory tasks requiring accuracy, dexterity, and reproducibility.

The laboratory associate must have a Bachelor's degree in chemistry or biology, and one to three years of related experience.

Laboratory associates should enjoy scientific and technical activities and should be able to make generalizations, evaluations, and decisions based on measurable or verifiable criteria. Situations requiring precise tolerances or standards are commonplace for the laboratory associate, and anyone planning to pursue this career should be well-prepared for the demands this job requires.

Technical Secretary

Technical secretaries provide word processing/secretarial support for research and development personnel, and perform office duties as required in the biotechnology company.

Technical secretaries who have two years of college and three to five years of typing/word processing experience, with the ability to type 60-80 words per minute, are preferred. Biotechnology companies prefer technical secretaries who have worked at least one year in a research or medical facility.

Technical secretaries should enjoy activities concerned with the communication of data, and tasks of a routine, concrete, and organized nature. Technical secretaries should be able to adapt to situations requiring precise tolerances or standards.

Data Entry Assistant

Data entry assistants, in the biotechnology company, select data processing methods, modify standard formulas, translate data, and are responsible for computer entry and for handling clinical studies and other research data.

Data entry assistants select the most economical and reliable combination of manual, mechanical, or electronic data processing methods and equipment consistent with data reduction requirements. They modify standard formulas to conform to the data processing methods selected, and translate data into numerical values, equations, flow charts, graphs, or other media.

Data entry assistants must have a college or technical degree, experience in computer technology, and at least two years experience in computer data handling, preferably with clinical studies.

Data entry assistants should enjoy pursuing scientific and technical activities and tasks that are carried on in relation to processes, machines, and techniques. Data entry assistants should be able to perform a variety of duties, often changing from one task to another of a different nature without losing efficiency. Toleration of work situations requiring precise tolerances or standards is also helpful.

Senior Instrumentation Technician

Senior instrumentation technicians are responsible for the repair and maintenance of all types of lab equipment used in the biotechnology company.

The senior instrumentation technician is required to have an AA in electronics or an equivalent degree, three to five years of experience, and good hands-on knowledge in the use of test equipment.

Senior instrumentation technicians should prefer activities dealing with things rather than people. They should like scientific and technical activities, and tasks carried on in relation to processes, machines, and techniques. Senior instrumentation technicians should be able to make generalizations, evaluations, and decisions based on measurable or verifiable criteria, and be able to tolerate situations requiring precise tolerances or standards.

Senior Maintenance Mechanic

Senior maintenance mechanics perform general building maintenance and related duties in the biotechnology company as needed. Under general direction and using power, hand, and welding tools, they install, maintain, repair, rebuild, and modify equipment and heavy mechanical lab equipment. Senior maintenance mechanics disassemble malfunctioning instruments to examine and test the mechanism and circuitry for defects. They repair heating, ventilation, air conditioning, and refrigeration equipment. Also, senior maintenance mechanics troubleshoot equipment in or out of control systems, and replace or repair defective parts. The job of reassembling instruments and testing those assemblies for conformance with specifications using instruments falls to senior maintenance mechanics. Inspecting biotechnology instruments periodically and making minor calibration adjustments to insure they function within specified standards are also the senior maintenance mechanic's responsibilities.

A high school diploma or the equivalent, plus five years journey-level mechanical maintenance and installation experience is required to be a senior maintenance mechanic.

Senior maintenance mechanics should prefer activities dealing with things rather than people, and should enjoy scientific and technical activities and tasks carried on in relation to processes, machines, and techniques. Senior maintenance mechanics should be able to make generalizations, evaluations, and decisions based on measurable or verifiable criteria, and deal with situations requiring precise tolerances or standards.

Laboratory Head

In the biotechnology company, the laboratory head directs and plans all laboratory projects; is in charge of the lab budget; provides reports; and supervises and directs senior scientists, research scientists, research assistants, and laboratory assistants. The laboratory head directs all biotechnology projects within a functional group, including planning with other departments, monitoring projected milestones, and assigning project personnel and other resources. He or she plans programs for research, product development, improvement of manufacturing processes, and the analysis and testing of liquids, substances, compounds, and gases. He or she also directs biotechnology programs, staff, and schedules; and assigns projects according to the field or fields of specialization of the lab personnel. In addition the laboratory head approves modifying formulas, standards, specifications, and processes; and coordinates research and analysis with manufacturing processes to assure acceptability of the biotechnology product or service. Providing regular updates, documentation, and communication of departmental results to the director and/or vice-president for scientific programs is yet another job duty.

The laboratory head must have completed a Ph.D. degree in biology or another natural science and four years of project leadership responsibilities, or a combination of education and related experience in rank with a Ph.D. degree. The ability to direct the completion of projects utilizing the resources of functional groups and in coordination with other departments is also required.

The laboratory head should be interested in scientific and technical activities, tasks carried on in relation to processes, machines and techniques, and activities resulting in prestige or the esteem of others. Laboratory heads should be able to accept responsibility for the direction, control, and planning of an activity; and to influence people in their opinions, attitudes, or judgments about ideas and things.

Research Scientist

Research scientists conduct research into the biological aspects of humans or other animals to develop new theories and facts—or to test, prove, or modify known theories of life systems—and to design life support apparatus.

Research scientists plan detailed procedures for defined biotechnology projects including time frames, milestones, methodological approaches, expected results, and necessary resources; participate in the conduct and implementation of defined projects; and interpret the scientific validity of the results. They also provide regular updates, documentation, and communication of biotechnology project results to senior scientists or laboratory heads. Research scientists, in addition, may supervise and direct research associates and lab assistants.

Research scientists are required to have a Ph.D. degree in biology or another natural science and one year of applicable experience, or an equivalent combination of education and related experience.

Research scientists should be interested in scientific and technical activities, and tasks that are carried on in relation to processes, machines, and techniques. Research scientists should be able to accept responsibility for the direction, control, and planning of an activity, and to make generalizations, evaluations, and decisions based on measurable or verifiable criteria.

Research Associate

Research associates prepare biological samples, compile information, and perform surgical procedures on animals. Research associates also prepare and analyze biological samples for laboratory procedures, and prepare reagents and chemical solutions. They assure that required reagents and solutions are available for each biotechnology project. Research associates also compile and maintain information from the results of the above analyses.

Research associates must have a bachelor's degree in biology or another natural science and one year experience, or a master's degree in biology or another natural science or an equivalent mix of education and experience in an applicable discipline. The ability to operate and understand appropriate laboratory equipment, instrumentation, and techniques is also required.

Research associates should prefer activities dealing with things rather than people, and should enjoy scientific and technical activities, and tasks of an abstract, creative nature.

Nuclear Medicine Technologist

Nuclear medicine technologists participate in biotechnology activities involving radioactive drugs used for the diagnosis and treatment of medical problems. They calculate and administer the correct dosage of drugs and are involved with the safety procedures required when using them.

Nuclear medicine technologists are required to have completed high school and at least two years of college in the natural sciences or a nuclear medical technology program, or a specialized program in the field.

Nuclear medicine technologists should enjoy scientific and technical activities, working for the presumed good of people, and activities resulting in tangible, productive satisfaction. Nuclear medicine technologists should be able to perform a variety of duties, often changing from one task to another of a different nature without losing efficiency, and make generalizations, evaluations, and decisions based on measurable or verifiable criteria.

Summary

The biotechnology industry is an integral part of the medical field. Biotechnology companies create, manufacture, and sell products and services that enhance and help the medical field. The products and developments mentioned in this chapter are some examples of what the biotechnology industries have provided to both the medical field and the consumer.

The biotechnology industry is creating new jobs. As this industry grows and diversifies, the need for more biotechnical personnel will increase. With each new discovery or product created, more people are needed to continue the research and production of the product. This field draws two disciplines together: the medical industry and the new biotechnology industry.

CHAPTER SIX

COMPUTER INDUSTRY

Computer Technology

Because the computer is increasingly used by consumers and small businesses in virtually all fields, the computer industry, despite a shake-up in 1985, will continue to expand—enormously! Computer technology is the foundation for the industry that encompasses the manufacturing of computer equipment (referred to as hardware), the development of software, and research on the application of artificial intelligence technologies. The manufacturing of computer hardware includes all the components, such as CPU (Central Processing Unit), and peripheral equipment such as disks, storage devices, tape drives, and printers. Computer software consists of the program or programs comprised of machine readable instructions that perform a logical sequence of functions, varying from maintaining bookkeeping entries to performing computer games. Artificial intelligence is the special computer software designed to mimic human reasoning. Computer hardware, computer software, and artificial intelligence make up the giant computer industry that has emerged over the past twenty years.

Computer Hardware

When big mainframe computers first became available for commercial use, only larger companies could afford to invest in the new technology. With the advent of small, less expensive minicomputers and microcomputers, the market for computers opened to private consumers and small businesses.

Small business owners are beginning to realize that computers can save time and money by freeing employees from tedious, repetitive tasks. Computers now store addresses, form letters, and make other documents easily accessible. Business use of computers also includes recording financial transactions, storing mailing lists, and assisting in long- and short-range planning. A single computer chip can store an extremely large amount of factual information, definitely a plus for businesses.

In the home, computers are used by private consumers to keep financial records, balance checkbooks, prepare the household budget, and play games. They will eventually enable consumers to access a wide range of information stored in commercial data bases. They will also be used for shopping at home, paying bills, and for many other household assists.

Computer hardware companies continually enhance and develop new hardware products. Enhanced products are often more advanced in competency power, less expensive, and easier to use. Computer hardware companies develop and manufacture a wide range of computer equipment, including desktop computers for engineering and scientific applications, and microcomputers for small businesses to fulfill both business and technical needs.

IBM is the largest computer manufacturing company in the world. Hewlett Packard is a fast-growing smaller company that manufactures computer hardware, develops software products, and markets and services its computer systems throughout the world. In 1983 Hewlett Packard had more than 72,000 employees, and manufacturing sites in 32 cities—21 in the United States and 10 abroad. HP also maintains approximately 200 sales and support offices and distributorships throughout 70 countries. Another smaller but rapidly growing manufacturer is Digital Equipment Corporation. IBM and several smaller companies (of which Hewlett Packard and Digital Equipment Corporation are but two) compete in this growing industry, for both customers and employees.

Computer Software

Software is a growing element in the computer industry in need of personnel at all stages of development, research, marketing, and sales. Software is the heart of the computer, the instructions that tell computers what to do. Without software, the computer would be useless—like a radio without circuitry or tubes.

There are three different sources developing computer software. First, computer hardware manufacturers develop operating systems, that is, software for people to make use of their computer hardware. Second, independent software firms develop commercial software packages such as Lotus 1, 2, 3, Pac-Man, etc. Third, individual contractors are hired to develop software products for specific applications.

Until recently, software was used primarily on mainframe computers. With the advent of personal computers, software is now developed and purchased by individual consumers. Software is being used for more and more applications, with thousands of different programs written and published every year. Such basic business applications as accounting, inventory control and payroll are available. New programs for construction contractors, such as Escomate, help contractors make bids, develop charts, do payroll, and even break down the costs of materials and labor. Software for the consumer maintains the family budget, does financial planning, and helps youngsters learn to read.

As the software industry grows, it continues to advance. The proliferation of software companies and the subsequent competition has given rise to unique, ever-improving programs in such diverse areas as word processing, general education, and elaborate games. Another improvement is the development of integrated software. Integrated software allows a person to perform several tasks with the same software program. According to *Time* magazine (April 16, 1984, page 62), "Someone using a computer to keep track of the family budget, for example, could classify his expenses into a variety of categories, [and] see how the budget might change if, say, interest rates went down and then look at a chart that represents how his money is being spent. All that can be done with just a few key strokes in less than a minute." Software is also being developed that will help link groups of personal computers to a central computer, allowing data to be freely exchanged among several personal computer users. Another major change, still on the drawing boards, will allow users to write computer commands in plain English, not in the confusing codes and symbols that are currently required.

The software now in use will seem very simplistic compared to the programs that will be developed in the future. As programmers learn and experiment, more programs will need to be developed. Future software programs will reflect this new knowledge. Research in artificial intelligence will continue to grow and the systems that are developed will enable the computer to act more as a thinking assistant than as an inanimate non-thinking machine.

Artificial Intelligence

Unlike the human brain, computers cannot reason, or understand and interpret instructions. Researchers are working to develop a programming language that will instruct a computer to simulate human reasoning. This simulation is called artificial intelligence (AI), and the computer systems that will produce AI are known as expert systems. The development of expert systems is extremely complex, and wide availability will take many years.

As researchers experiment with AI, their programs improve and become more effective. The first expert systems were taught to understand grammatical units from sentences. Later, realizing that this was not teaching the computer to think, researchers taught the systems to extract meanings from sentences themselves. Researchers were, and still are, trying to teach the computer to think or to appear as if it is thinking. Research has progressed through the years, and the computer has been taught to paraphrase meanings, translate, and draw inferences from sentences. The computer has also been taught to tie the meaning of one sentence to that of another by interpreting the link between the sentences. This progressive research and development of AI led researchers to the conclusion that they need to develop a system with a dynamic memory, one that changes every time it understands a story. As

Roger C. Schank and Peter G. Childers say in their article in *Computer World,* "We are trying to develop a system that is capable of building up an increasing number of memories about different situations it has experienced, which it then can use in understanding even more stories and experiences."

As explained above, AI systems perform well above and beyond the conventional computer. Take this example from the field of geology: In the time of World War I, geologists reasoned that a rich deposit of molybdenum ore was buried deep under Mount Tolman in eastern Washington. Sixty years later this metal had still not been found. The AI system, PROSPECTOR, an electronic geologist, located the molybdenum because SRI International, its developer, made it possible for the computer to react like a geologist. PROSPECTOR was programmed to duplicate human reasoning. Nine geologists were interviewed about how they locate minerals from what they observe in the field. This accumulated knowledge was programmed into PROSPECTOR. The information, reduced to a set of rules, was combined with a large data base of geological information and then made available to the program. PROSPECTOR asked questions about the Mount Tolman area and, after computing the answers, was able to determine the location of the ore. The program was successful because it could accumulate and sort an enormous amount of data. This is just one way in which expert systems can be developed and utilized.

AI systems will be developed in the future for a multitude of uses. They will be used by doctors and consumers to diagnose diseases. This application is possible because most of a doctor's skill entails applying diagnostic rules to evidence presented by a patient. Such a program will enable patients to consult an information system prior to seeing a doctor. AI systems may also one day be used in manufacturing, as the brains for the robots. These robots will assist engineers and managers in such different tasks as supervising orders and inventories, designing products, and coordinating production. Artificial intelligence can also be used to develop totally automatic personal vehicles, guided by matching intelligence. These cars will be powered by electricity drawn from internal batteries while in towns and on minor roads and from a main supply on the highways. These cars will not only be quiet and pollution free, but also free from human error. The possibilities for AI research and development are almost limitless.

Many corporations, such as ITT, Litton, General Electric, and Hughes Aircraft are setting up AI laboratories because they see its possibilities for developing into a multibillion dollar industry. There are five different subareas in AI:

1. Expert Systems
2. Natural Language Software
3. Computer Aided Instruction
4. Visual Recognition
5. Voice Recognition

Each of these areas is rapidly growing. It will not be long before computers will reason almost as well as people, learn from experience, and communicate in human or natural languages. Researchers in artificial intelligence are making such computers a reality and their potential use is vast. AI systems will be used in all sectors of society, public and private. They will play a significant role in our future.

Careers in the Computer Industry

The next section of this chapter describes the key occupations that exist in computer hardware, software, and artificial intelligence sections of the computer industry. This section will also describe the training and preparation required, pay scales, and the abilities and interests typically associated with people successful at these occupations.

Designer

Designers in the computer industry research and develop new computer hardware products. Designers analyze data to determine the feasibility of a product and confer with research personnel to clarify or resolve problems, system layouts and detailed drawings, and schematics. After feasibility is determined, designers develop a detailed description of the product. Designers also plan and develop experimental test programs and analyze test results to determine if the new computer hardware product meets functional and performance specifications.

A minimum of a bachelor's degree in computer science or engineering is required, but a master's degree in engineering is preferred. The salary range for designers falls between $15,000 and $25,000 annually.

Computer hardware designers should gravitate toward activities dealing with the communication of scientific and technical data. Accepting responsibility for the direction, control, and planning of an activity; making generalizations, evaluations, and decisions based on measurable or verifiable criteria; and dealing with people beyond giving and receiving instructions are capabilities required of computer hardware designers.

Engineer

Computer product design engineers work with designers to develop new computer hardware products. An engineer as-assists the designer in the development of specifications and also supervises other engineers. Design engineers spend most of their time actually building the hardware product.

The engineer is required to have a bachelor's degree in computer science or engineering. The salary range falls between $18,000 and $35,000 annually.

Computer design engineers should be interested in activities dealing with the communication of scientific and technical data.

Project Manager

A project manager working in the computer hardware industry basically does what the job title states, that is, managing hardware development projects. It is the project manager's responsibility that tasks performed by designers, engineers, and technical people are accomplished within budget, on time, and as specified. The project manager organizes the project team, and needs to know the status of each hardware development task during each phase of the project.

The project manager is required to have a bachelor's degree in computer science, but an MBA is preferred. The salary ranges from $20,000 to $40,000 annually.

The project manager should enjoy working with things and objects, and scientific and technical activities at work.

Manager of Research & Development

The research and development (R & D) manager oversees the entire research and development operation that results in the design and development of new computer hardware. The R & D manager directs and coordinates activities concerned with new concepts, ideas, basic data on, and application for a computer hardware manufacturer's new products and services.

The research and development manager in the computer hardware industry must have a bachelor's degree in computer science or in a related engineering field. The salary ranges from $30,000 to $50,000 annually.

The R & D manager should enjoy business contact with people and activities resulting in prestige or the esteem of fellow workers.

Independent Consultant

Consultants to the computer hardware industry work independently and may serve several clients. Independent consultants confer with clients to define their needs, then conduct studies and surveys to obtain data, and analyze that data to advise or recommend solutions, utilizing their knowledge of computer hardware theory, principles, and technology. Consultants are engaged for specific assignments and continue to work for the hardware manufacturer for the duration of the assignment. Consultants may work with the hardware designers at one company, advising them of new or different ways to create desired hardware, or may assist technical writers in developing the final documentation or operating instruction manuals.

Independent computer hardware consultants should have a bachelor's degree in computer science or engineering, and experience in computer hardware and logic design. The salary ranges from $25,000 to $50,000 annually.

Independent computer hardware consultants should enjoy scientific and technical work activities resulting in the prestige or esteem of clients, and consulting engagements that deal with processes, machines, and techniques.

Technician

Computer manufacturing technicians help fabricate, test, analyze, and adjust prototypes of computer hardware under development, following specifications developed by a designer or engineer. Technicians also assemble electrical components and analyze test results, and repair and adjust prototype computer hardware.

A technician is required to have a high school diploma and a two-year associate of arts degree in a technical area. The starting salary ranges between $10,000 and $20,000 annually.

Computer hardware manufacturing technicians should like work that includes things and objects, and scientific and technical activities carried on in relation to processes, machines, and techniques. Technicians usually enjoy activities resulting in tangible, productive satisfaction and can adapt to a variety of duties, often changing from one task to another without losing efficiency or composure.

Fabricator

Fabricators in the computer hardware manufacturing industry assemble the hardware components and products.

Fabricators secure components onto a circuit board, then wire and test the new hardware components.

A computer hardware component fabricator must have a high school diploma and trade school credits. The salary ranges from $10,000 to $20,000 annually.

Fabricators should enjoy work dealing with things and objects, and scientific and technical activities carried on in relation to processes, machines, and techniques. Fabricators tend to like work resulting in tangible and productive satisfaction and can perform a variety of duties, often changing from one to another without losing efficiency or composure.

Line Supervisor

A line supervisor in a computer hardware manufacturing company applies a knowledge of the product design, the assembly process, and the capacities of line workers to guarantee the efficiency of assembly-line production. The line supervisor also orders the materials necessary to meet production goals, and examines assembled computer components to determine if they meet design specifications. Finally, the line supervisor trains new workers.

The line supervisor is required to have a high school diploma. The salary ranges from $15,000 to $25,000 annually.

A line supervisor in the computer hardware manufacturing industry should enjoy business contact with people and activities carried on in relation to processes, machines, and techniques that result in prestige or the esteem of other workers.

Technical Writer

Technical writers in the computer industry write, edit, and proofread the operating instructions and user manuals that support the use of the software and hardware produced by a computer hardware manufacturer. Technical writers may write user manuals to explain how to use a software operating system, instructions for installing an entire computer system, catalogs, and related technical and administrative publications.

The technical writer is required to have a bachelor's degree in computer science and to have taken classes in journalism and English. The salary ranges from $15,000 to $25,000 annually.

The technical writer should enjoy work tasks that include the communication of data, and scientific and technical activities.

Course Instructor

The course instructor prepares and conducts in-house technical training programs for employees in the computer industry. These educational programs are very important to a company because they cover all aspects of company products. The course instructor selects or develops such teaching aids as training handbooks, demonstration models, multimedia visual aides, and reference works. Finally the course instructor tests employees to measure their learning progress and to evaluate the effectiveness of the training program.

A course instructor in the computer industry is required to have a bachelor's degree in computer science with credits in teaching. A master's degree is preferred. The salary ranges from $20,000 to $35,000 annually.

The course instructor should be interested in activities concerned with the communication of data, and business contact with fellow workers. An instructor should be able to accept responsibility for the direction, control, and planning of a work activity.

Education and Training Administrator

The education and training administrator develops training programs to educate a computer manufacturer's employees. It is important to the company that the education and training administrator teach employees the company's special way of conducting business. After conferring with the company's management to determine training needs, the education and training administrator formulates training policies and schedules utilizing knowledge of identified training needs, company production processes, and business systems or changes in computer products, procedures, or services. The education and training administrator also trains course instructors and supervisory personnel in effective techniques for training the employees of the computer hardware manufacturer.

The education and training administrator must have a minimum of a bachelor's degree in computer science, and have taken business and teaching classes as well. The salary ranges from $20,000 to $35,000 annually.

An education and training administrator in the computer industry should enjoy work activities concerned with the communication of data, and business contact with people. An administrator should be able to take responsibility for the direction, control, and planning of a work activity. Influencing employees in their opinions, attitudes, and judgments about the computer company is an important duty of the education and training administrator.

Software Designer

Software designers may be employed by a computer manufacturer or a software company. A software designer creates the operating systems that allow people to interface with the computer, and the software products that perform specific functions or applications (such as accounting systems). Software designers use schematics and write specifications to describe the software product being designed.

The software designer is required to have a bachelor's degree in computer science, and a master's degree is preferable. The salary ranges between $15,000 and $25,000 annually.

Systems Analyst

Systems analysts are employed by computer hardware manufacturers to assist customers, referred to as users, with the implementation of new systems and applications. They may formulate mathematical or simulation models of a user's problem for solution by computers. Systems analysts working for a software company develop and design new computer systems based on the type of businesses for which the company develops products. Systems analysts work with programmers to design a software system, and perform validation and testing of the system to insure validity and reliability. Systems analysts are responsible for the design of the entire computer system or product.

A systems analyst is required to have a bachelor's degree in computer science. The salary range falls between $20,000 and $35,000 annually.

Systems analysts should enjoy scientific technical work tasks, and activities that relate to processes, machines, and techniques. Systems analysts should be able to perform a variety of duties, often changing from one to another without losing efficiency.

Manager of Planning

The manager of planning in the computer industry plans and prepares production schedules for the manufacturing of the software and/or hardware products. The manager of planning also plans the sequence of fabrication, assembly, installation, and other manufacturing operations for the guidance of computer production.

The manager of planning must have at least a bachelor's degree in computer science. The salary ranges between $30,000 and $50,000 annually.

The manager of planning should prefer scientific and technical work activities, and tasks that are related to processes, machines, and techniques.

Documentation Librarian

Documentation librarians within the computer industry maintain files of all written materials pertaining to hardware and software systems developed in the company. The librarians organize, file, and issue operating manuals and other written materials.

A documentation librarian is required to have a high school diploma and have taken some computer classes. The salary range falls between $10,000 and $20,000 annually.

Documentation librarians working in the computer industry should be able to manage a variety of duties, often changing from one to another without losing efficiency or composure. Dealing with employees beyond giving and receiving instructions is another necessary skill.

Business Applications Programmer

A business applications computer programmer develops programs to perform such tasks as accounting, payroll, and database management using a computer. A business programmer analyzes all or part of the work flow charts or diagrams developed by a systems analyst that represent business problems. By applying a knowledge of computer capabilities and the subject matter, a programmer develops a sequence of program steps onto detailed logical flow charts in symbolic form that represents the flow of data to be processed by the computer system. The flow charts describe all data input, output, and arithmetic and logical operations involved. The business programmer then converts the detailed logical flow chart into instructions written in a language processable by computers. Finally, the business programmer tests the completed program to check the program's operating efficiency.

A business programmer working in the computer industry is required to have at least a high school degree with an associate of arts degree in programming. A bachelor's degree in computer science is preferred. The salary ranges from $12,000 to $25,000 annually.

A business programmer should enjoy scientific and technical work activities, and activities that are carried on in relation to processes, machines, and techniques.

Scientific Applications Programmer

A scientific applications programmer has job duties similar to the business programmer but must also convert scientific, engineering, and other technical problem formulations to a format processable by computer.

The scientific applications programmer is required to have a bachelor's degree in engineering. The salary range is from $18,000 to $30,000 annually.

The scientific programmer should enjoy scientific and technical work tasks and activities of an abstract, creative nature.

Contract Programmer

Like consultants, contract programmers are engaged for a specific programming task when hired by a software company or computer hardware manufacturer. Contract programmers design, write, and document programs under contract for a software or hardware firm. Contract programmers either work for a firm that specializes in providing programmers to businesses, or are independent contractors working for themselves.

A contract programmer usually has a bachelor's degree in computer science, and is paid either an hourly rate or a fixed price for the engagement.

The contract programmer should enjoy scientific and technical work activities, and tasks carried on in relation to processes, machines, and techniques. Contract programmers should be able to accept responsibility for the direction, control, and planning of an activity; make generalizations, evaluations, and decisions based on measurable or verifiable criteria; and tolerate situations requiring the precise attainment of set limits, tolerances, and standards.

Data-Base Administrator

A data-base administrator works in the computer industry or a corporation designing, creating, and operating new data bases. The data-base programmer's primary job duties are the design and testing of computer files organized into a data-base structure. The data-base structure is an accessible base of information available to users for information. The data-base programmer develops the programs that build and maintain the data base.

A data-base programmer should have an MBA along with a bachelor's degree in computer science. The salary range falls between $25,000 and $35,000 annually.

A data-base programmer should enjoy scientific and technical work activities, and tasks carried on in relation to processes, machines, and techniques.

Telecommunications Network Analyst

A telecommunications network analyst in the computer industry designs and builds systems (networks) that transfer, via telephone lines, data and information between computers and endwires. These networks provide a vast resource of information for end-users. The network analyst is responsible for the programming required to develop the network system, and the installation and testing of the telecommunications network.

The telecommunications network analyst must have a bachelor's degree in computer science with advanced classes in communications and network theory. The salary ranges from $20,000 to $35,000 annually.

Telecommunications network analysts should be interested in work activities that are scientific and technical, and related to processes, machines, and techniques.

Chief Programmer

The chief programmer's primary responsibility, when employed in the computer industry, is supervising a team or teams of programmers working on a software development project or projects. The chief programmer also plans, schedules, and directs the preparation of programs to process business data and solve business-oriented problems. Assigning, coordinating, and reviewing the work of programming personnel is the responsibility of the chief programmer, as is the training of subordinates in programming and coding.

A chief programmer is required to have a bachelor's degree in computer science. The salary ranges from $25,000 to $35,000 annually.

The chief programmer should be interested in work activities concerned with the communication of data, and enjoy scientific and technical activities that are carried on in relation to processes, machines, and techniques and activities resulting in prestige or the esteem of fellow workers.

Manager-Systems Development

A manager of systems development within the computer industry plans, directs, and coordinates activities of designated systems development projects to insure that specified goals and objectives are accomplished in accordance with prescribed priorities, time, and budget considerations. The systems manager also confers with the systems staff to outline project plans, designate personnel responsible for all phases of the project, and establish the team members' scope of authority.

A systems development manager must have a bachelor's degree in computer science and an MBA. The salary ranges from $30,000 to $50,000 annually.

The manager of systems development should gravitate toward activities dealing with things and objects; scientific and technical tasks carried on in relation to processes, machines, and techniques; and activities resulting in prestige or the esteem of fellow employees.

Support or Maintenance Programmer

The support or maintenance programmer working in the computer industry takes an already existing computer program and corrects it as required or updates it as changes or revisions are needed. The support or maintenance programmer is also responsible for correcting, reviewing, and updating the user manual.

A support or maintenance programmer is required to have a bachelor's degree in computer science. The starting salary ranges from $12,000 to $20,000 annually.

The support or maintenance programmer should like scientific and technical programming activities that are related to processes and techniques. Support or maintenance programmers accept responsibility for the direction, control, and planning of programming tasks.

Systems Programmer

Systems programmers employed by a computer manufacturer write and test specialized computer software referred to as operating systems. Operating systems software controls the interface process that allows a computer to use application programs.

A systems programmer must have a bachelor's degree in computer science. The salary range is $20,000 to $35,000 annually.

Systems programmers should gravitate toward scientific and technical work activities, and activities that are carried on in relation to processes, machines, and techniques.

Manager-System Support

The system support manager working for a computer manufacturer oversees the activities of the systems programmers. The system support manager plans, directs, and coordinates the activities of designated projects on the development and maintenance of operating systems software to insure that project goals are accomplished within prescribed priorities, time, and budget expectations.

The system support manager must have a bachelor's degree in computer science. The salary ranges from $25,000 to $40,000 annually.

The system support manager should enjoy managing scientific and technical activities; managing activities carried on in relation to processes, machines, and techniques; and work activities that result in prestige and the esteem of fellow workers.

Computer Service Technician

Computer technicians service and repair computers for customers of computer manufacturers. Technicians also analyze the technical requirements of the customer (whether a big business or a single person), and install and maintain equipment. Computer service technicians first consult with the customer to plan the layout of the computer room, then direct the installation of the computer system according to manufacturer's specifications. Finally, computer service technicians operate the system to demonstrate it, and to train the customer's employees on how to use the computer equipment.

A computer technician is required to have a high school diploma with an associate of arts degree in computer servicing. The starting salary ranges from $10,000 to $20,000 annually.

Computer technicians should enjoy computer service activities involving contact with people and work activities resulting in tangible, productive satisfaction.

Computer Equipment Technical Specialist

Computer equipment technical specialists working for a computer manufacturer specialize in the installation and maintenance of a specific type of computer related equipment. Technical specialists are required to know all there is to know about the equipment they are responsible for, and to maintain that equipment.

The computer equipment technical specialist is required to have a high school diploma and an associate of arts degree in computer service. The salary ranges from $15,000 to $25,000 annually.

Computer equipment technical specialists should be interested in equipment service activities involving contact with people, and activities concerned with the communication of data.

Service Manager

The service manager is in charge of all the technicians and field engineers employed by a computer manufacturer. The service manager is responsible for scheduling, hiring, problem-solving, and generally keeping customers satisfied with the performance of the computer, and the field engineer's and technicians' services.

The service manager is required to have a bachelor's degree in computer science. The salary ranges between $20,000 and $40,000 annually.

The service manager should enjoy managing scientific and technical activities, and activities resulting in prestige or the esteem of fellow workers.

Sales Representative

Sales representatives sell the products of the computer industry. Sales representatives must have a thorough knowledge of the computer or software system that they represent since they must often give formal presentations to the prospective customer.

The sales representative must have a bachelor's degree in marketing and have taken classes in computer science. The salary range is variable because a representative's compensation is usually a combination of salary and commission. The average base salary is $25,000 to $35,000 annually.

Sales representatives should enjoy sales activities involving direct contact with people, and sales activities concerned with the communication of data.

Sales/Marketing Engineer

Sales/marketing engineers develop new sales campaigns, research market conditions, and plan marketing strategies for companies in the computer industry. Sales/marketing engineers review a variety of economic reports and forecasts, and create a marketing plan or sales campaign.

A sales/marketing engineer is required to have a bachelor's degree in marketing or sales engineering as well as course credit in computer science classes. The salary ranges from $20,000 to $35,000 annually.

Sales/marketing engineers should be interested in marketing activities concerned with the communication of scientific and technical data. A marketing engineer should be able to accept responsibility for the direction, control, and planning of a sales activity; to influence people through sales campaigns in their opinions, attitudes, and judgments about ideas or things; and to make generalizations, evaluations, and decisions based on measurable or verifiable criteria.

Sales/Marketing Manager

The sales/marketing manager within the computer industry directs staffing, training, and the performance evaluauations of sales representatives to develop and control the sales program. The sales/marketing manager coordinates sales distribution by establishing sales territories, a quota, and goals. The sales/marketing manager also reviews market analysis to determine customer needs and volume potential. He or she works with the marketing engineer to develop sales campaigns that accommodate the goals of the computer company.

A sales/marketing manager is required to have a minimum of a bachelor's degree in marketing or sales engineering, with classes in computer science. The salary range falls between $30,000 and $50,000 annually.

The sales/marketing manager should enjoy management activities involving business contact with people, sales activities concerned with the communication of data, and activities resulting in prestige or the esteem of employees and customers. Performing a variety of management duties without losing efficiency, and accepting responsibility for the direction, control, and planning of a sales activity is necessary for a sales/marketing manager.

CHAPTER SEVEN

ENERGY INDUSTRY

The Energy Industry Today

In 1879 Thomas Alva Edison and his staff invented the first practical incandescent electrical lamp. With that invention the electrical revolution was born and the energy industry began. Today the energy industry is still growing, and a new energy revolution is under way.

Electric power comes from nine primary energy resources: water, geothermal, wind, solar power, biomass, oil, gas, nuclear power, and coal. But not all energy companies obtain electricity from a variety of energy resources. These energy resources can be divided into two groups. Conventional power systems use oil, gas, nuclear power, and coal to produce electricity. Alternate and renewable power systems use such natural energy sources as water, wind, geothermal, solar, and biomass resources to produce electricity. The energy revolution under way today was triggered by federal legislation passed to encourage the development of alternate and renewable power systems that lessen our dependence on oil. This change in the use of power systems is expanding the job market in the energy industry.

The energy industry is experimenting with natural energy resources that do not create irreversible pollution (such as smog) when they create energy. For example, wind-created electricity may create noise and visual pollution from use of the wind turbine, but these are not permanent pollutions like smog. This period is one of transition from dependence on fossil fuels to one of learning how to develop products that keep our world habitable.

As these changes take place in energy companies, new occupations are created and more people are needed in existing occupations. The renewable and alternative energy resources being developed and tested to produce electric power are wind, solar, geothermal, and biomass. The development and use of these resources is expanding the energy industry and in turn creating more jobs.

Wind turbines, once they are built, can produce electricity without depleting resources or causing air pollution. Located where winds are strong and persistent, a wind turbine generator (WTG) is used with other energy sources. The WTG design promises future cost savings and performance improvements. By the year 2000, as much as ten percent of California's and many other states' electrical energy could be produced from renewable wind resources.

Two new technologies produce electricity from the sun. Solar thermal uses sun power to produce heat energy for generating electricity, and solar photovoltaic converts sunlight directly into electrical energy. Both need no fossil fuel, and are clean and quiet.

Geothermal resources, in the form of hot water or steam locked beneath the earth's surface, are used to generate electricity. Where the earth's crust has heat reservoirs, wells can tap this energy. Two decades were needed to develop technology to use the hot, salty, corrosive water for generating electricity.

Biomass/recycled waste is becoming an energy resource as cities and industry cope with the challenge of waste disposal. Wood, which once supplied 90 percent of America's energy needs, is experiencing a rebirth as an energy producer. Some energy companies are using wood waste to produce a clean, combustible fuel gas in a downdraft wood gasifier.

These energy resource products not only create more jobs, but do not create pollution and are cost-effective.

Cogeneration plants produce another type of energy. Cogeneration facilities produce two types of energy from a single fuel, an ideal conservation measure. Many private companies create their own cogeneration plants. These private companies are not allowed to compete with the established energy companies. Therefore the government requires the private companies to obtain special permits to build cogeneration plants that produce more than 50 megawatts (one megawatt produces power for 500 homes) of power. Often private companies will build cogeneration plants that produce more than they need, so that they can sell the excess power to the energy companies. This is cost-effective for private companies because they now have a less expensive supply of electricity and are making money by selling the excess electricity to energy companies. This is also good for the energy companies because it reduces their need to expand their plants.

All of these new products and the consumer's constant need for increasing energy supplies means that the energy industry will continue to expand and be a source of many job opportunities in the future.

Careers in the Energy Industry

This next section will explain how to prepare for a career in the energy industry, the key occupations, the wages paid, and the work requirements for these occupations. There are three major areas of careers in the energy industry: Maintenance/Operations positions, Administrative/Professional positions, and Skilled/Journeyman positions.

Maintenance/Operations Positions

Testman Helper   Testman helpers inspect, repair, and maintain single phase watt-hour and demand meters. Inspection entails the observation of electrical and mechanical components by observing and recording readings of test apparatus meters. Repairs include the reconditioning and replacement of defective parts within the meters. Testman helpers also clean, paint, and remove minor defects from meter boxes, protective covers, and mounting brackets.

This position usually requires a high school education and at least 30 units of college level courses covering basic electricity. A working knowledge of the principles of trigonometry is also helpful. The salary range for this occupation is between $18,000 and $21,600 per year.

Testman helpers should be interested in activities working with things that relate to processes, machines, and techniques that result in tangible, productive satisfaction. Testman helpers should be able to make evaluations and decisions based on measurable or verifiable criteria, and to tolerate situations requiring the precise attainment of tolerances and standards.

Nuclear Plant Equipment Operator

Nuclear plant equipment operators inspect, test, clean, perform maintenance, and manually operate electrical, mechanical, and hydraulic equipment. Nuclear plant equipment operators inspect meters, indicators, and gauges to detect abnormal fluctuations. The accuracy of flowmeters, pressure gauges, temperature indicators, controllers, radiation counters, ore detectors, and other recordings are tested by the nuclear plant operator using special test equipment. They also use indicating or controlling instruments to locate defective components in the system. Nuclear plant equipment operators must think clearly and act quickly, effectively, and reasonably under normal emergency conditions. A daily log, recording unusual events and changes in conditions of operation, must be maintained.

A high school education is the minimum requirement for a nuclear plant equipment operator and the salary range is $22,800 to $26,400 per year.

Nuclear plant equipment operators should prefer working with things rather than people, and should enjoy activities of a scientific and technical nature, and work that is carried on in relation to processes, machines, and techniques.

Administrative/Professional Positions

Nuclear Engineer Nuclear engineers prepare safety analysis reports, monitor plant performance, conduct research and development, and write technical reports concerning the operation of the nuclear plant. Nuclear engineers prepare safety analysis reports for commercial nuclear plants, and prepare and maintain plant safety or environmental technical specifications. Plant performance is monitored to insure efficient functioning and conformance with safety specifications, regulations, and laws. Nuclear engineers also conduct research into problems of nuclear energy systems, and plan and conduct nuclear research to discover facts or to test, prove, or modify known nuclear theories concerning the release, control, and utilization of nuclear energy. Nuclear engineers monitor the testing, operation, and maintenance of nuclear reactors, and design and develop reactor cores, radiation shielding, and associated instrumentation and control mechanisms. Finally, nuclear engineers prepare technical reports concerning research and development activities and inspectional functions.

A bachelor's or master's degree in nuclear, mechanical, or electrical engineering is required. The starting salary ranges from $22,800 to $32,400 per year, but with experience and proper qualifications can increase to $36,000 to $54,000 per year.

To be a nuclear engineer one should be interested in scientific and technical activities that are of an abstract, creative nature, and like tasks concerned with processes, machines, and techniques. Nuclear engineers should be able to accept responsibility for the direction, control, and planning of an activity; deal with generalizations, evaluations, and decisions based on measurable or verifiable criteria; and work in situations requiring precise tolerances and standards.

Electrical Engineer Electrical engineers plan and translate consumer energy requirements, direct electrical equipment usage, estimate costs, and inspect finished products. While translating the energy requirements, electrical engineers plan generation and transmission resources to meet the needs of consumers, and analyze the expected results of conservation and local management programs. To meet consumers' needs for electrical energy, electrical engineers coordinate operation and maintenance activities to insure the optimum utilization of power system facilities.

An electrical engineer may design generating plants, transmission and distribution lines, and receiving and distribution stations. In doing so, he or she estimates the necessary labor, material, construction, and equipment costs. The electrical engineer then directs the preparation of specific types of equipment and materials to be used in construction and equipment installation of the new facilities, and inspects the completed installation for conformance with design and equipment specifications and safety standards.

A bachelor's degree in electrical engineering is required and the starting salary ranges between $22,800 and $32,400 per year. With experience and proper qualifications, an electrical engineer's salary can increase to $36,000 to $54,000 per year.

Electrical engineers should be interested in scientific and technical activities and like tasks that are carried on in relation to processes, machines, and techniques.

Civil Engineer

Civil engineers plan, design, and direct the construction and maintenance of power facilities. Civil engineers also perform technical research to develop solutions to problems discovered during maintenance of facilities.

A bachelor's degree in civil engineering is required and the starting salary range is $22,800 to $32,400 per year. With experience and proper qualifications, the yearly salary can increase to $36,000 to $54,000.

Civil engineers should find activities of a scientific, technical, abstract, and creative nature interesting and attractive.

Environmental Engineer

Environmental engineers perform engineering tasks associated with generation station operations, and explore how generating stations affect the environment.

A bachelor's degree in mechanical engineering, chemical engineering, or environmental engineering is required. The starting salary ranges from $22,800 to $32,400 per year, but with experience and the proper qualifications can increase to $36,000 to $54,000 per year.

Start-Up Engineer

Start-up engineers conduct and prepare tests and plan, coordinate, and evaluate electric power systems. Start-up engineers prepare preoperational and initial start-up phase testing, conduct the tests, and verify and evaluate test results. Planning enables start-up engineers to provide orderly development and to improve the operating efficiency of the electric power system. Start-up engineers coordinate the scheduling, conducting, and analysis of such special studies as commercial and residential developments in surrounding territories, population estimates, and advantages of and facilities for interconnections with other power systems. To meet the requirements of new, increased, or future loads, or to improve an already existing system, start-up engineers evaluate, analyze, and recommend additional facilities.

A bachelor's or master's degree in nuclear, mechanical, or electrical engineering is required. With six years of experience, the salary ranges from $36,000 to $54,000 per year.

Start-up engineers should be interested in activities of a scientific and technical nature, and activities that are carried on in relation to processes, machines, and techniques. They should prefer activities dealing with things rather than people.

Nuclear Assistant Shift Supervisor

The nuclear assistant shift supervisor supervises and coordinates control room operations. These operations include installing, adjusting, repairing, and maintaining electrical and mechanical parts of the equipment. The nuclear assistant shift supervisor also trains workers in repair and maintenance procedures.

One must have at least a high school education, and have completed courses in math, physics, chemistry, and English. The nuclear assistant shift supervisor must complete a training program and obtain a Senior Reactor Operator License. The salary ranges from $26,400 to $39,600 per year.

The nuclear assistant shift supervisor should enjoy activities involving contact with people, activities that are carried on in relation to processes, machines, and techniques, and tasks resulting in prestige and the esteem of others.

Telecommunications Engineer

Telecommunications engineers are in charge of designing point-to-point microwave systems, telephone systems, PBX trunking systems, switching networks, and UHF/VHF local base stations.

A telecommunications engineer must have a bachelor of science degree in electrical engineering. The starting salary ranges between $26,400 and $39,600 per year, but with experience and the proper qualifications can rise to $36,000 to $55,200 per year.

Telecommunications engineers should prefer activities dealing with things rather than people, and should enjoy activities of a scientific and technical nature and activities that are carried on in relation to processes, machines, and techniques.

Mechanical Engineer

Mechanical engineers perform tasks associated with generating station operation and maintenance. Mechanical engineers also design mechanical or electromechanical products or systems, and direct and coordinate operation and repair activities. In addition, mechanical engineers direct and coordinate construction and installation activities to insure that products and systems conform with engineering design and customer specifications. These activities are coordinated by mechanical engineers to obtain optimum utilization of machines and systems.

A mechanical engineer must have a bachelor's degree in mechanical or chemical engineering. The starting salary ranges from $22,800 to $32,400 per year, but with experience and proper qualifications can increase to $36,000 to $55,200 per year.

Mechanical engineers should prefer activities dealing with things rather than people, and like activities of a scientific and technical nature, and tasks carried on in relation to processes, machines, and techniques.

Chemical Engineer

Chemical engineers monitor water treatment practices in power plants, and they analyze, develop, and design new procedures and equipment for the plant. Chemical engineers also perform tests and take measurements throughout all stages of production to determine the degree of control over such variables as temperature, density, specific gravity, and pressure. They also apply principles of chemical engineering to solve environmental problems.

Chemical engineers are responsible for the analysis of new testing equipment, chemicals, and operating procedures. They also analyze equipment and machinery functions to reduce processing time and cost. Chemical engineers design equipment and develop processes for manufacturing chemicals and related products utilizing principles and technology of chemistry, physics, mathematics, engineering, and related physical and natural sciences. Chemical engineers also develop new water treatment procedures working with new technologies.

A bachelor's or master's degree in chemical engineering is required. The starting salary ranges from $26,400 to $38,400 per year, but with experience and the proper qualifications can increase from $36,000 to $55,200 per year.

Chemical engineers should be interested in activities of a scientific and technical nature, and activities that are carried on in relation to processes, machines, and techniques. Chemical engineers should be able to perform a variety of duties, often changing from one task to another of a different nature without losing efficiency. They also accept responsibility for the direction, control, and planning of activities.

Plant Engineer

The plant engineer provides engineering support, directs and coordinates personnel and activities in the plant, establishes standards and policies for plant operation, and tests equipment. The plant engineer maintains and supports plant performance and operating efficiency through engineering assistance. Engineering support is also supplied during start-up, for preventative maintenance, and during scheduled power outages during plant shutdowns. The plant engineer directs and coordinates, through engineering and supervisory personnel, activities concerned with the design, construction, and maintenance of equipment and machinery in the plant. In addition, the plant engineer establishes standards and policies for pollution control, testing, operating procedures, inspection, and maintenance of equipment in accordance with engineering principles and safety regulations. He or she also oversees, directly or through subordinates, the maintenance of plant buildings. Preparing bid sheets and contracts for construction and facility acquisition, and testing the newly installed machines and equipment according to contract specifications are also responsibilities of the plant engineer.

Being a plant engineer requires a bachelor's or master's degree in mechanical, chemical, or electrical engineering. It would be advisable for the plant engineer to have experience in fossil-fuel steam generation plant performance, and familiarity with boilers, turbine generators, pumping, heat exchangers, and process control equipment. The starting salary range is $26,400 to $38,400 per year, but with experience and the proper qualifications can increase to $38,400 to $58,000 per year.

The plant engineer should prefer activities dealing with things rather than people, and like activities of a scientific and technical nature, and tasks carried on in relation to processes, machines, and techniques.

Industrial Engineer

Industrial engineers review and evaluate the general productivity of corporate departments or operating units with regard to manpower utilization, operating efficiency, and resource utilization. Industrial engineers accomplish this by establishing a way of measuring the volume of work done, called work measurement programs. Further, they make sample observations of work to develop standards of manpower utilization. The responsibilities of industrial engineers also include measurement of the utilization of manpower and machines against established standards and the development of work simplification programs.

A four-year college degree in engineering or the equivalent, with an emphasis in industrial engineering or production engineering is required. It is advisable for the industrial engineer to be exposed to utility operations, and to be a registered engineer with exposure to business administration. The salary range is $26,400 to $38,400 per year.

Industrial engineers should enjoy activities of a scientific and technical nature, and activities that are carried on in relation to processes, machines, and techniques.

Quality Assurance Engineer

The quality assurance engineer performs actions that assure the plant's smooth running, develops procedures, directs workers, and compiles material to write training manuals for sessions on quality control activities.

The quality assurance engineer performs actions necessary to provide adequate assurance that a structure system or component will perform satisfactorily in service, and that construction and operational activities will conform with prescribed requirements. To provide control and measurement of the quality of items to predetermined requirements is one of the actions performed. The quality assurance engineer develops and initiates methods and procedures for inspection, testing, and evaluation. He or she also directs workers engaged in measuring and testing products, and tabulating quality and reliability data.

A bachelor's degree in electrical or mechanical engineering is required. The salary ranges from $31,200 to $46,800 per year or from $36,000 to $54,000 per year, depending on the engineer's experience and qualifications.

Quality assurance engineers should enjoy activities of a scientific and technical nature and tasks that are carried on in relation to processes, machines, and techniques.

Valuation Engineer

Valuation engineers direct and participate in the development, compilation, and analysis of data related to depreciation accruals, estimated net salvage values, and useful lives of company facilities for accounting and regulatory reporting purposes. They are also involved in such specialized, non-routine studies as the annual cost of alternatives in the relicensing of hydraulic facilities. Further, valuation engineers prepare depreciation or trend chart exhibits for governmental rate proceedings.

Four years of college or the equivalent, with an emphasis in engineering math or statistics is required. It is also preferred that the valuation engineer be a professional engineer familiar with computer programming techniques. The salary ranges from $31,200 to $46,800 per year.

Valuation engineers should prefer activities dealing with things rather than people, be interested in activities of a scientific and technical nature, and like tasks that are abstract and creative. Valuation engineers should be able to accept responsibility for the direction, control, and planning of an activity, and be able to deal with people at a level higher than merely giving and receiving instructions.

Controls Designer

Controls designers perform all phases of design work including the complete design of components and systems for energy companies. The phases of design work include developing layouts from engineering specifications, preparing the calculations necessary to perform design work, selecting materials and equipment, and preparing bills of materials.

A controls designer must have five years of design and drafting, or three years of design with a bachelor's degree in architecture. The starting salary range is $26,400 to $38,400 per year and can increase with experience and proper qualifications to $31,200 to $46,800 per year.

Controls designers should be interested in activities concerned with the communication of data and activities of a scientific and technical nature. Controls designers should be able to accept responsibility for the direction, control, and planning of an activity, and deal with people at a level higher than merely giving and receiving instruction.

Electrical Designer

Electrical designers perform all phases of design work including developing layouts from engineering specifications, preparing calculations necessary for performing design work, selecting materials and equipment, and preparing bills of materials.

Five years of design and drafting experience or three years of design with a bachelor's degree in architecture is required. The starting salary ranges from $26,400 to $38,400 per year, and depending on the experience and qualifications, can increase to $31,200 to $46,800 per year.

Electrical designers should enjoy activities of a scientific and technical nature, and like tasks that are carried on in relation to processes, machines, and techniques.

Drafting Technician

Using engineering data and standard references, drafting technicians draft technical drawings—detailed multiview drawings of machines and subassemblies including specifications concerning gear ratios, bearing loads, and the direction of moving parts. Drafting technicians gather and compute data from marked prints, rough or detailed sketches, field notes, and verbal instructions. They also check finished drawings and the computations of others. Another responsibility of drafting technicians is to analyze engineering sketches, specifications, and related data and drawings to determine such design factors as the size, shape, and arrangement of parts.

A high school education and experience in chemical, oil refinery, or power plant work with a special emphasis on piping or instrumentation is required. The salary range falls between $20,400 and $24,000 per year.

Drafting technicians should prefer activities dealing with things rather than people, and enjoy activities of a scientific and technical nature that are carried out in relation to processes, machines, and techniques.

Health Physics Engineer (Nuclear)

The health physics engineer protects plant and lab personnel from radiation hazards by performing an advisory role, directing research, and consulting with scientific personnel. As an advisor to the radiation protection program, the health physics engineer participates in audits and reviews, and devises and directs research, training, and monitoring programs to protect plant and laboratory personnel from radiation hazards. Research is also conducted to develop inspection standards, radiation exposure limits, safe work methods, and decontamination procedures. Surrounding areas are also tested to insure that radiation is not in excess of permissible standards. To determine that equipment or plant design conforms to health physics standards for protection of personnel, the health physics engineer consults with scientific personnel regarding new experiments.

A bachelor's or master's degree in health physics or nuclear engineering is required. The starting salary ranges from $22,800 to $32,400 per year; with experience and the proper qualifications, the salary range can reach $36,000 to $55,200 per year.

To be a health physics engineer one should be interested in activities of a scientific and technical nature, and activities that are carried on in relation to processes, machines, and techniques.

Research Scientist

Research scientists plan and conduct assignments requiring judgment in independent evaluation, selection and substantial modification, and application of standard or advanced scientific principles, theories, and concepts.

A bachelor's degree in a scientific discipline or engineering, and five years experience as an associate research scientist, or three to four years equivalent experience with a master's degree in designing and conducting independent scientific studies is advisable. The starting salary ranges from $26,400 to $39,600 per year; with experience and proper qualifications, it can increase to $38,400 to $58,800 per year.

The research scientist should find activities concerned with the communication of data and activities of a scientific and technical nature interesting.

Biologist

Biologists develop, conduct, and analyze studies of the marine environment; or operate and maintain the energy company's fisheries, providing technical assistance in ecological matters affecting hydro-generation.

Biologists must have a bachelor's degree in biology or zoology, and four years of experience in marine biology or two years of field study with fish and wildlife. The starting salary ranges from $26,400 to $39,600 per year, and with experience and proper qualifications, can increase to $36,000 to $55,200 per year.

Biologists should enjoy activities of a scientific and technical nature. Making generalizations, evaluations, and decisions based on measurable or verifiable criteria; and coping with situations requiring the precise attainment of set limits, tolerances, or standards are necessary skills for biologists.

Special Agent

The special agent conducts investigations and physical security surveys for the company, and investigates cases of fraud, theft, burglary, robbery, sabotage, and property damage. The special agent also conducts physical security surveys of company facilities and specialized investigations for the law and claims departments in the company. Investigations of fraud regarding the metering of electrical energy are also the special agent's responsibility.

A bachelor's degree in industrial security and/or police science, and a master's degree in industrial security or police security is advisable. The special agent must be creative, diplomatic, factual, accurate, and available 24 hours per day, 365 days per year. The salary ranges from $31,200 to $48,000 per year.

To be a special agent you should enjoy activities involving business contact with people and tasks concerned with the communication of data. Performing well under stress when confronted with emergency, critical, unusual, or dangerous situations in which working speed and sustained attention are vital to success are mandatory qualifications of a secret agent. Dealing with people at a level higher than merely giving and receiving instructions is also necessary.

Skilled/Journeyman Positions

Test Technician — Test technicians test, inspect, repair, and adjust relays, meters, and associated devices for the protection and operation of lines, generators, motors, transformers, and regulators for voltages.

Two years of college-level electronics is required and the salary ranges from $24,000 to $32,400 per year.

To be a test technician one should prefer activities dealing with things rather than people, and like tasks that are carried on in relation to processes, machines, and techniques.

Laboratory Assistant — Laboratory assistants perform routine chemical analyses relating to chemical investigations, assist in the set-up of equipment, and perform standard chemical analyses and tests on chemical cleaning solvents, oils, fuels, deposits, and ion exchange resins.

A minimum of one year of college chemistry is required to be a laboratory assistant and the salary ranges between $19,800 and $25,200 per year.

Laboratory assistants should be interested in activities dealing with things rather than people, activities of a scientific and technical nature, and activities that are carried on in relation to processes, machines, and techniques. Laboratory assistants should be able to perform a variety of duties, often changing from one task to another of a different nature without losing efficiency, and make generalizations, evaluations, and decisions based on measurable or verifiable criteria.

Nuclear Chemical Technician

Nuclear chemical technicians collect samples of water, solids, or gas and perform standard chemical treatments or adjustments on them to maintain established control limits. Nuclear chemical technicians perform the necessary environmental monitoring surveys to determine radiation levels. They also calibrate and service chemical and radiation detection instrumentation. Conducting chemical and physical laboratory tests and making qualitative and quantitative analyses of materials, liquids, and gases for such purposes as research, the development of new products and materials, processing and production methods, quality control, maintenance of health and safety standards, and other tests involving experimental, theoretical, or practical applications of chemistry and related sciences are the major responsibilities of nuclear chemical technicians.

A high school education is required and the salary range falls between $27,600 and $30,000 per year for nuclear chemical technicians.

Nuclear chemical technicians should take interest in activities dealing with things rather than people, activities of a scientific and technical nature, and activities that are carried on in relation to processes, machines, and techniques.

Instrument Technician

Instrument technicians inspect, service, repair, and install such indicating, recording, and automatic control instruments as flow meters, pressure gauges, thermometers, controllers regulators, chlorimeters, and similar components. Using such test equipment as portable pressure gauges manometers, potentiometers, conductivity bridges, signal generators, and oscilloscopes are some of the responsibilities of instrument technicians.

Two years of college-level electronics and the ability to work from logic and matrix diagrams is required. A good working knowledge of electrical and pressure instruments and devices as used in complex analog and digital control systems and their component parts is also necessary. The salary is $30,000 per year.

Instrument technicians should enjoy activities dealing with things rather than people, activities of a scientific and technical nature, and activities that are carried on in relation to processes, machines, and techniques.

Communication Technician

Communication technicians construct, operate, and maintain microwave and multiplex systems, mobile radio systems, and power line, open wire, and cable carrier systems.

A college electronics course and an FCC second class license are required. The salary ranges from $24,000 to $30,000 per year.

Communication technicians should be interested in activities of a scientific and technical nature, activities that are carried on in relation to processes, machines, and techniques, and tasks resulting in tangible, productive satisfaction.

Chemical Technician

Chemical technicians perform routine chemical analyses relating to chemical investigations, and assist in the set-up of equipment. Chemical technicians also collect samples of chemical cleaning solvents, oils, fuels, deposits, and ion exchange resins and perform standard chemical analysis and tests on them.

Two years of college-level chemistry is required to be a chemical technician. The salary range is $26,400 to $28,800 per year.

The chemical technician should like activities dealing with things rather than people, activities of a scientific and technical nature, and tasks that are carried on in relation to processes, machines, and techniques.

Nuclear Instrument Technician

Nuclear instrument technicians repair, calibrate, and test instrumentation including reactor plant controls and protective equipment, turbine plant controls, radiation counting and detaching instrumentation, and nuclear flux counting and detaching equipment. Nuclear instrument technicians then report test findings; decontaminate, clean, and lubricate instruments and tools; and keep mechanical and electrical diagrams and drawings up-to-date reflecting the changes and alterations.

A high school education and a technical training or study in nuclear instrumentation is required of those wanting to be nuclear instrument technicians. The salary is $30,000 per year.

Nuclear instrument technicians should be interested in activities dealing with things rather than people, activities of a scientific and technical nature, and tasks that are carried on in relation to processes, machines, and techniques. Nuclear instrument technicians should be able to work in situations requiring precise tolerances and standards, and make generalizations, evaluations, and decisions based on measurable or verifiable criteria.

Steam Electrician

Steam electricians install, inspect, and test electrical equipment, and notify plant personnel of necessary equipment downtime to maintain uninterrupted services.

Steam electricians install and inspect electrical components in generating plants, and dismantle, inspect, repair, replace, reassemble, and adjust generators, transformers, power circuit breakers, motors, and other similar equipment. Testing defective equipment, using voltmeters, ammeters, and related electrical testing apparatus to determine the cause of malfunction or failure, is also a responsibility falling to steam electricians.

Steam electricians must have a high school education. The salary is $28,800 per year.

Steam electricians should enjoy activities dealing with things rather than people, activities that are carried on in relation to processes, machines, and techniques, and activities resulting in tangible, productive satisfaction.

Construction Electrician

Construction electricians work at high elevations, construct substation and transmission apparatus, and supervise apprentices.

A high school education and such job-related courses as general science and shop mathematics are required of those wanting to be construction electricians. The starting salary is $28,800 per year.

Construction electricians should excel in activities involving business contact with people, and activities carried on in relation to processes, machines, and techniques. Many of the job's activities result in prestige and the esteem of others.

Steam Welder

Steam welders are involved in all phases of welding. Steam welders build up and weld metal parts by means of oxyacetylene or electric welding apparatus, fabricate metals, and repair cracked and broken parts. Steam welders cut metal with oxyacetylene cutting machines, grind welds to conform with contours, and weld parts of superheaters, boilers, and other high pressure vessels. Selecting torch, torch tip, filter rod, and flux according to welding chart specifications, or type and thickness of metal; connecting regulator valves and hoses to oxygen and fueling gas cylinders; and using a welding torch are some of steam welders' job responsibilities. Steam welders also turn regular valves to activate the flow of gases, light torch, and, based on knowledge of gas-welding techniques, adjust the gas mixture and pressure to obtain the desired flame.

Steam welders must have a high school education. The salary is $28,800 per year.

To be a steam welder you should prefer activities dealing with things rather than people, and like tasks that are carried on in relation to processes, machines, and techniques.

Lineman/Splicer

Linemen/splicers (hereafter referred to as linemen) climb poles and structures, and perform required duties in an elevated position. Linemen also use live line tools to work on energized lines, and drive and operate hole digging, pole setting, wire stringing, and other types of equipment used in constructing and maintaining power lines.

Linemen must have a high school education. The salary range is from $28,800 to $30,000 per year.

A lineman should enjoy activities dealing with things rather than people, activities carried on in relation to processes, machines, and techniques that result in tangible, productive satisfaction.

Health Physics Technician

Health physics technicians collect and perform radionuclide analysis of plant and environmental samples to protect plant employees, the public, and the environment. Health physics technicians assess radiation facility contamination levels, post control signs and barriers, and measure the intensity and identify the type of radiation in working areas using special devices. Health physics technicians monitor and control worker radiation exposures, and inform supervisors when individual exposures and area radiation levels approach the maximum permissible limit. Recommending work stoppage in unsafe areas, posting warning signs, roping off contaminated areas, and preparing radiation waste shipments are additional job responsibilities. Health physics technicians also collect air samples to determine the airborne concentration of radioactivity. They also collect and analyze monitoring equipment worn by personnel, such as film badges and pocket detection chambers, to measure individual exposure to radiation.

Health physics technicians must have a high school education supplemented by a specialized study in nuclear physics. The salary is $27,600 per year.

A health physics technician should take interest in activities of a scientific and technical nature and activities that are carried on in relation to processes, machines, and techniques.

APPENDIX

Résumés, Application Forms, Cover Letters, and Interviews

by Neale Baxter

You might see a hurdle to leap over, or a hoop to jump through. Or a barrier to knock down. That is how many people think of résumés, application forms, cover letters, and interviews. But you do not have to think of them that way. They are not ways to keep you from a job; they are ways for you to show an employer what you know and what you can do. After all, you are going to get a job. It is just a question of which one.

Employers want to hire people who can do the job. To learn who these people are, they use résumés, application forms, written tests, performance tests, medical examinations, and interviews. You can use each of these different evaluation procedures to your advantage. You might not be able to make a silk purse out of a sow's ear, but at least you can show what a good ear you have.

This article is reprinted from *Occupational Outlook Quarterly,* spring 1987, volume 31, number 1, pp. 17-23.

Creating Effective Résumés and Application Forms

Résumés and application forms are two ways to achieve the same goal: To give the employer written evidence of your qualifications. When creating a résumé or completing an application form, you need two different kinds of information: facts about yourself and facts about the job you want. With this information in hand, you can present the facts about yourself in terms of the job. You have more freedom with a résumé—you can put your best points first and avoid blanks. But, even on application forms, you can describe your qualifications in terms of the job's duties.

Know thyself Begin by assembling information about yourself. Some items appear on virtually every résumé or application form, including the following:

- Current address and phone number—if you are rarely at home during business hours, try to give the phone number of a friend or relative who will take messages for you.
- Job sought or career goal.
- Experience (paid and volunteer)—date of employment, name and full address of the employer, job title, starting and finishing salary, and reason for leaving (moving, returning to school, and seeking a better position are among the readily accepted reasons).
- Education—the school's name, the city in which it is located, the years you attended it, the diploma or certificate you earned, and the course of studies you pursued.
- Other qualifications—hobbies, organizations you belong to, honors you have received, and leadership positions you have held.
- Office machines, tools, equipment you have used, and skills that you possess.

Other information, such as your Social Security number, is often asked for on application forms but is rarely presented on résumés. Application forms might also ask for a record of past addresses and for information that you would rather not reveal, such as a record of convictions. If asked for such information, you must be honest. Honesty does not, however, require that you reveal disabilities that do not affect your overall qualifications for a job.

Know thy job

Next, gather specific information about the jobs you are applying for. You need to know the pay range (so you can make their top your bottom), education and experience usually required, hours and shifts usually worked. Most importantly, you need to know the job duties (so that you can describe your experience in terms of those duties). Study the job description. Some job announcements, especially those issued by a government, even have a checklist that assigns a numerical weight to different qualifications so that you can be certain as to which is the most important; looking at such announcements will give you an idea of what employers look for even if you do not wish to apply for a government job. If the announcement or ad is vague, call the employer to learn what is sought.

Once you have the information you need, you can prepare a résumé. You may need to prepare more than one master résumé if you are going to look for different kinds of jobs. Otherwise, your résumé will not fit the job you seek.

Two kinds of résumés

The way you arrange your résumé depends on how well your experience seems to prepare you for the position you want. Basically, you can either describe your most recent job first and work backwards (reverse chronology) or group similar skills together. No matter which format you use, the following advice applies generally.

- Use specifics. A vague description of your duties will make only a vague impression.

- Identify accomplishments. If you headed a project, improved productivity, reduced costs, increased membership, or achieved some other goal, say so.

- Type your résumé, using a standard typeface.

(Printed résumés are becoming more common, but employers do not indicate a preference for them.)

- Keep the length down to two pages at the most.
- Remember your mother's advice not to say anything if you cannot say something nice. Leave all embarrassing or negative information off the résumé—but be ready to deal with it in a positive fashion at the interview.
- Proofread the master copy carefully.
- Have someone else proofread the master copy carefully.
- Have a third person proofread the master copy carefully.
- Use the best quality photocopying machine and good white or off-white paper.

The following information appears on almost every résumé.

- Name.
- Phone number at which you can be reached or receive messages.
- Address.
- Job or career sought.
- References—often just a statement that references are available suffices. If your references are likely to be known by the person who reads the résumé, however, their names are worth listing.
- Experience.
- Education.
- Special talents.
- Personal information—height, weight, marital status, physical condition. Although this information appears on virtually every sample résumé I have ever seen, it is not important according to recruiters. In fact, employers are prohibited by law from asking for some of it. If some of this information is directly job related—the height and weight of a bouncer is important to a

disco owner, for example—list it. Otherwise, save space and put in more information about your skills.

Reverse chronology is the easiest method to use. It is also the least effective because it makes when you did something more important than what you can do. It is an especially poor format if you have gaps in your work history, if the job you seek is very different from the job you currently hold, or if you are just entering the job market. About the only time you would want to use such a résumé is when you have progressed up a clearly defined career ladder and want to move up a rung.

Résumés that are not chronological may be called functional, analytical, skill oriented, creative, or some other name. The differences are less important than the similarity, which is that all stress what you can do. The advantage to a potential employer—and, therefore, to your job campaign—should be obvious. The employer can see immediately how you will fit the job. This format also has advantages for many job hunters because it camouflages gaps in paid employment and avoids giving prominence to irrelevant jobs.

You begin writing a functional résumé by determining the skills the employer is looking for. Again, study the job description for this information. Next, review your experience and education to see when you demonstrated the ability sought. Then prepare the résumé itself, putting first the information that relates most obviously to the job. The result will be a résumé with headings such as "Engineering," "Computer Languages," "Communications Skills," or "Design Experience." These headings will have much more impact than the dates that you would use on a chronological résumé.

Fit yourself to a form

Some large employers, such as fast food restaurants and government agencies, make more use of application forms than of résumés. The forms suit the style of large organizations because people find information more quickly if it always appears in the same place. However, creating a résumé before filling out an application form will still benefit you. You can use the résumé when you send a letter inquiring about a position. You can submit a résumé even if an application is required; it will spotlight your qualifications. And the information on the résumé will serve as a handy reference if you must fill out an application form quickly. Application forms are really just résumés in disguise anyway. No matter how rigid the form appears to be, you can still use it to show why you are the person for the job being filled.

At first glance, application forms seem to give a job hunter no leeway. The forms certainly do not have the flexibility that a résumé does, but you can still use them to your best advantage. Remember that the attitude of the person reading the form is not, "Let's find out why this person is unqualified," but, "Maybe this is the person we want." Use all the parts of the form—experience blocks, education blocks, and others—to show that that person is you.

Here's some general advice on completing application forms.

- Request two copies of the form. If only one is provided, photocopy it before you make a mark on it. You'll need more than one copy to prepare rough drafts.

- Read the whole form before you start completing it.

- Prepare a master copy if the same form is used by several divisions within the same company or organization. Do not put the specific job applied for, date, and signature on the master copy. Fill in that information on the photocopies as you submit them.

- Type the form if possible. If it has lots of little lines that are hard to type within, type the information on a piece of blank paper that will fit in the space, paste the paper over the form, and photocopy the finished product. Such a procedure results in a much neater, easier to read page.

- Leave no blanks; enter n/a (for "not applicable") when the information requested does not apply to you; this tells people checking the form that you did not simply skip the question.

- Carry a résumé and a copy of other frequently asked information (such as previous addresses) with you when visiting potential employers in case you must fill out an application on the spot. Whenever possible, however, fill the form out at home and mail it in with a résumé and a cover letter that point up your strengths.

Writing Intriguing Cover Letters

You will need a cover letter whenever you send a résumé or application form to a potential employer. The letter should capture the employer's attention, show why you are writing, indicate why your employment will benefit the company, and ask for an interview. The kind of specific information that must be included in a letter means that each must be written individually. Each letter must also be typed perfectly, which may present a problem. Word processing equipment helps. Frequently only the address, first paragraph, and specifics concerning an interview will vary. These items are easily changed on word processing equipment and memory typewriters. If you do not have access to such equipment, you might be able to rent it. Or you might be able to have your letters typed by a résumé or employment services company listed in the yellow pages. Be sure you know the full cost of such a service before agreeing to use one.

Let's go through a letter point by point.

Salutation — Each letter should be addressed by name to the person you want to talk with. That person is the one who can hire you. This is almost certainly not someone in the personnel department, and it is probably not a department head either. It is most likely to be the person who will actually supervise you once you start work. Call the company to make sure you have the right name. And spell it correctly.

Opening — The opening should appeal to the reader. Cover letters are sales letters. Sales are made after you capture a person's attention. You capture the reader's attention most easily by talking about the company rather than yourself. Mention projects under development, recent awards, or favorable comments recently published about the company. You can find such information in the business press, including the business section of local newspapers and the many magazines that are devoted to particular industries. If you are answering an ad, you may mention it. If someone suggested that you write, use their name (with permission, of course).

Body The body of the letter gives a brief description of your qualifications and refers to the résumé, where your sales campaign can continue.

Closing You cannot have what you do not ask for. At the end of the letter, request an interview. Suggest a time and state that you will confirm the appointment. Use a standard complimentary close, such as "Sincerely yours," leave three or four lines for your signature, and type your name. I would type my phone number under my name; this recommendation is not usually made, although phone numbers are found on most letterheads. The alternative is to place the phone number in the body of the letter, but it will be more difficult to find there should the reader wish to call you.

Triumphing on Tests and at Interviews

A man with a violin case stood on a subway platform in The Bronx. He asked a conductor, "How do you get to Carnegie Hall?" The conductor replied, "Practice! Practice! Practice!"

Tests That old joke holds good advice for people preparing for employment tests or interviews. The tests given to job applicants fall into four categories: General aptitude tests, practical tests, tests of physical agility, and medical examinations. You can practice for the first three. If the fourth is required, learn as soon as possible what the disqualifying conditions are, then have your physician examine you for them so that you do not spend years training for a job that you will not be allowed to hold.

To practice for a test, you must learn what the test is. Once again, you must know what job you want to apply for and for whom you want to work in order to find out what tests, if any, are required. Government agencies, which frequently rely on tests, will often provide a sample of the test they use. These samples can be helpful even if an employer uses a different test. Copies of standard government tests are usually available at the library.

If you practice beforehand, you'll be better prepared and less nervous on the day of the test. That will put you ahead of the competition. You will also improve your performance by following this advice:

- Make a list of what you will need at the test center, including a pencil; check it before leaving the house.

- Get a good night's sleep.

- Be at the test center early—at least 15 minutes early.

- Read the instructions carefully; make sure they do not differ from the samples you practiced with.

- Generally, speed counts; do not linger over difficult questions.

- Learn if guessing is penalized. Most tests are scored by counting up the right answers; guessing is all to the good. Some tests are scored by counting the right answers and deducting partial credit for wrong answers; blind guessing will lose you points—but if you can eliminate two wrong choices, a guess might still pay off.

Interviews

For many of us, interviews are the most fearsome part of finding a job. But they are also our best chance to show an employer our qualifications. Interviews are far more flexible than application forms or tests. Use that flexibility to your advantage. As with tests, you can reduce your anxiety and improve your performance by preparing for your interviews ahead of time.

Begin by considering what interviewers want to know. You represent a risk to the employer. A hiring mistake is expensive in terms of lost productivity, wasted training money, and the cost of finding a replacement. To lessen the risk, interviewers try to select people who are highly motivated, understand what the job entails, and show that their background has prepared them for it.

You show that you are highly motivated by learning about the company before the interview, by dressing appropriately, and by being well mannered—which means that you greet the interviewer by name, you do not chew gum or smoke, you listen attentively, and you thank the interviewer at the end of the session. You also show motivation by expressing interest in the job at the end of the interview.

You show that you understand what the job entails and that you can perform it when you explain how your qualifications prepare you for specific duties as described in the company's job listing and when you ask intelligent questions about the nature of the work and the training provided new workers.

One of the best ways to prepare for an interview is to have some practice sessions with a friend or two. Here is a list of some of the most commonly asked questions to get you started.

- Why did you apply for this job?
- What do you know about this job or company?
- Why should I hire you?
- What would you do if . . . (usually filled in with a work-related crisis)?
- How would you describe yourself?
- What would you like to tell me about yourself?
- What are your major strengths?
- What are your major weaknesses?
- What type of work do you like to do best?
- What are your interests outside work?
- What type of work do you like to do least?
- What accomplishment gave you the greatest satisfaction?
- What was your worst mistake?

- What would you change in your past life?
- What courses did you like best or least in school?
- What did you like best or least about your last job?
- Why did you leave your last job?
- Why were you fired?
- How does your education or experience relate to this job?
- What are your goals?
- How do you plan to reach them?
- What do you hope to be doing in 5 years? 10?
- What salary do you expect?

Many jobhunting books available at libraries discuss ways to answer these questions. Essentially, your strategy should be to concentrate on the job and your ability to do it no matter what the question seems to be asking. If asked for a strength, mention something job related. If asked for a weakness, mention a job-related strength (you work too hard, you worry too much about details, you always have to see the big picture). If asked about a disability or a specific negative factor in your past—a criminal record, a failure in school, being fired—be prepared to stress what you learned from the experience, how you have overcome the shortcoming, and how you are now in a position to do a better job.

So far, only the interviewer's questions have been discussed. But an interview will be a two-way conversation. You really do need to learn more about the position to find out if you want the job. Given how frustrating it is to look for a job, you do not want to take just any position only to learn after 2 weeks that you cannot stand the place and have to look for another job right away. Here are some questions for you to ask the interviewer.

- What would a day on this job be like?
- Whom would I report to? May I meet this person?
- Would I supervise anyone? May I meet them.?

- How important is this job to the company?
- What training programs are offered?
- What advancement opportunities are offered?
- Why did the last person leave this job?
- What is that person doing now?
- What is the greatest challenge of this position?
- What plans does the company have with regard to . . . ? (Mention some development of which you have read or heard.)
- Is the company growing?

After you ask such questions, listen to the interviewer's answers and then, if at all possible, point to something in your education or experience related to it. You might notice that questions about salary and fringe benefits are not included in the above list. Your focus at a first interview should be the company and what you will do for it, not what it will pay you. The salary range will often be given in the ad or position announcement, and information on the usual fringe benefits will be available from the personnel department. Once you have been offered a position, you can negotiate the salary. The jobhunting guides available in bookstores and at the library give many more hints on this subject.

At the end of the interview, you should know what the next step will be: Whether you should contact the interviewer again, whether you should provide more information, whether more interviews must be conducted, and when a final decision will be reached. Try to end on a positive note by reaffirming your interest in the position and pointing out why you will be a good choice to fill it.

Immediately after the interview, make notes of what went well and what you would like to improve. To show your interest in the position, send a followup letter to the interviewer, providing further information on some point raised in the interview and thanking the interviewer once again. Remember, someone is going to hire you; it might be the person you just talked to.

Chronological Résumé

Allison Springs
15 Hilton House
College de l'Art Libre
Smallville, CO 77717

(888) 736-3550

Job sought:

Education

| September 1984 to June 1988 | College de l'Art Libre College Lane Smallville, CO 77717 | Vice President, Junior Class (raised $15,000 for junior project) Member College Service Club (2 years) Swim Team (4 years) Harvest Celebration Director Major: Political science with courses in economics and accounting |

Experience

Period employed	Employer	Job title and duties
January 1988 to present 10 hours per week	McCall, McCrow, and McCow 980 Main Street Westrow, CO 77718 Supervisor: Jan Eagelli	Research assistant: Conducted research on legal and other matters for members of the firm.
September 1987 to December 1987 10 hours per week	Department of Public Assistance State of Colorado 226 Park Street Smallville, CO 77717 Supervisor: James Fish	Claims interviewer: Interviewed clients to determine their eligibility for various assistance programs. Directed them to special administrators when appropriate.
Summers 1981-1986	Shilo Pool 46 Waterway Shilo, NE 77777 Supervisor: Leander Neptune	Lifeguard: Insured safety of patrons by seeing that rules were obeyed, testing chemical content of the water, and inspecting mechanical equipment.

Recommendations available on request

Functional Résumé

Allison Springs
15 Hilton House
College de l'Art Libre
Smallville, CO 77717

(888) 736-3550

Job sought: Food Industry Sales Representative

Skills, education, and experience

Negotiating skills: My participation in student government has developed my negotiating skills, enabling me both to persuade others of the advantages to them of a different position and to reach a compromise between people who wish to pursue different goals.

Promotional skills: The effective use of posters, displays, and other visual aids contributed greatly to my successful campaign for class office (Junior Class Vice President), committee projects, and fund raising efforts (which netted $15,000 for the junior class project).

Skill working with people: All the jobs I have had involve working closely with people on many different levels. As Vice President of the Junior Class, I balanced the concerns of different groups in order to reach a common goal. As a claims interviewer with a state public assistance agency, I dealt with people under very trying circumstances. As a research assistant with a law firm, I worked with both lawyers and clerical workers. And as a lifeguard (5 summers), I learned how to manage groups. In addition, my work with the state and the law office has made me familiar with organizational procedures.

Chronology

September 1984 to present	Attended College de l'Art Libre in Smallville, Colorado. Will earn a Bachelor of Arts degree in political science. Elected Vice President of the Junior Class, managed successful fund drive, directed Harvest Celebration Committee, served on many other committees, and earned 33 percent of my college expenses.
January 1988 to present	Work as research assistant for the law office of McCall, McCrow, and McCow, 980 Main Street, Westrow, Colorado 77718. Supervisor: Jan Eagelli (666) 654-3211
September 1987 to December 1987	Served as claims interviewer intern for the Department of Public Assistance of the State of Colorado, 226 Park Street, Smallville, Colorado 77717. Supervisor: James Fish (666) 777-7717.
1981-1986	Worked as lifeguard during the summer at the Shilo Pool, 46 Waterway, Shilo, Nebraska 77777.

Recommendations available on request

Functional Résumé

Allison Springs
15 Hilton House
College de l'Art Libre
Smallville, CO 77717

(888) 736-3550

Job sought: Hotel Management Trainee

Skills, education, and experience

Working with people: All the jobs I have had involve working closely with a large variety of people on many different levels. As Vice President of the Junior Class, I balanced the concerns of different groups in order to reach a common goal. As a claims interviewer with a state public assistance agency, I dealt with people under very trying circumstances. As a research assistant with a law firm, I worked with both lawyers and clerical workers. And as a lifeguard (5 summers), I learned how to manage groups of people.

Effective communication: My campaign for class office, committee projects, and fund raising efforts (which netted $15,000 for the junior class project), relied on effective communication in both oral and written presentations.

Organization and management: My participation in student government has developed my organizational and management skills. In addition, my work with the state government and a law office has made me familiar with organizational procedures.

Chronology

September 1984 to present	Attended College de l'Art Libre in Smallville, Colorado. Will earn a Bachelor of Arts degree in political science. Elected Vice President of the Junior Class, managed successful fund drive, directed Harvest Celebration Committee, served on many other committees, and earned 33 percent of my college expenses.
January 1988 to present	Work as research assistant for the law office of McCall, McCrow, and McCow, 980 Main Street, Westrow, Colorado 77718. Supervisor: Jan Eagelli (666) 654-3211
September 1987 to December 1987	Worked as claims interviewer intern for the Department of Public Assistance of the State of Colorado, 226 Park Street, Smallville, Colorado 77717. Supervisor: James Fish (666) 777-7717.
1981-1986	Worked as lifeguard during the summers at the Shilo Pool, 46 Waterway, Shilo, Nebraska 77777.

Recommendations available on request

Cover Letter

15 Hilton House
College de l'Art Libre
Smallville, CO 77717
March 18, 1988

Ms. Collette Recruiter
Rest Easy Hotels
1500 Suite Street
Megapolis, SD 99999

Dear Ms. Recruiter:

The Rest Easy Hotels always served as landmarks for me when I traveled through this country and Europe. I would like to contribute to their growth, especially their new chain, the Suite Rest Hotels that feature reception rooms for every guest. I have had many jobs working with people and have always enjoyed this aspect of my experience. Knowing its importance to your company, I believe I would be an asset to the Rest Easy Hotels.

During the week of March 31, I will be visiting Megapolis and would like to speak with you concerning your training program for hotel managers. I will call your secretary to confirm an appointment.

The enclosed résumé outlines my education and experience.

Sincerely yours,

Allison Springs

Allison Springs
(888) 736-3550

VGM CAREER BOOKS

OPPORTUNITIES IN

Available in both paperback and hardbound editions

Accounting Careers
Acting Careers
Advertising Careers
Agriculture Careers
Airline Careers
Animal and Pet Care
Appraising Valuation Science
Architecture
Automotive Service
Banking
Beauty Culture
Biological Sciences
Book Publishing Careers
Broadcasting Careers
Building Construction Trades
Business Communication Careers
Business Management
Cable Television
Carpentry Careers
Chemical Engineering
Chemistry Careers
Child Care Careers
Chiropractic Health Care
Civil Engineering Careers
Commercial Art and Graphic Design
Computer Aided Design and Computer Aided Mfg.
Computer Maintenance Careers
Computer Science Careers
Counseling & Development
Crafts Careers
Dance
Data Processing Careers
Dental Care
Drafting Careers
Electrical Trades
Electronic and Electrical Engineering
Energy Careers
Engineering Technology Careers
Environmental Careers
Fashion Careers
Federal Government Careers
Film Careers
Financial Careers
Fire Protection Services
Fitness Careers
Food Services
Foreign Language Careers
Forestry Careers
Gerontology Careers
Government Service
Graphic Communications

Health and Medical Careers
High Tech Careers
Home Economics Careers
Hospital Administration
Hotel & Motel Management
Industrial Design
Insurance Careers
Interior Design
International Business
Journalism Careers
Landscape Architecture
Laser Technology
Law Careers
Law Enforcement and Criminal Justice
Library and Information Science
Machine Trades
Magazine Publishing Careers
Management
Marine & Maritime Careers
Marketing Careers
Materials Science
Mechanical Engineering
Microelectronics
Modeling Careers
Music Careers
Nursing Careers
Nutrition Careers
Occupational Therapy Careers
Office Occupations
Opticianry
Optometry
Packaging Science
Paralegal Careers
Paramedical Careers
Part-time & Summer Jobs
Personnel Management
Pharmacy Careers
Photography
Physical Therapy Careers
Plumbing & Pipe Fitting
Podiatric Medicine
Printing Careers
Psychiatry
Psychology
Public Health Careers
Public Relations Careers
Real Estate
Recreation and Leisure
Refrigeration and Air Conditioning
Religious Service
Retailing
Robotics Careers
Sales Careers

Sales & Marketing
Secretarial Careers
Securities Industry
Social Work Careers
Speech-Language Pathology Careers
Sports & Athletics
Sports Medicine
State and Local Government
Teaching Careers
Technical Communications
Telecommunications
Television and Video Careers
Theatrical Design & Production
Transportation Careers
Travel Careers
Veterinary Medicine Careers
Vocational and Technical Careers
Word Processing
Writing Careers
Your Own Service Business

CAREERS IN

Accounting
Business
Communications
Computers
Health Care
Science

CAREER DIRECTORIES

Careers Encyclopedia
Occupational Outlook Handbook

CAREER PLANNING

How to Get and Get Ahead On Your First Job
How to Get People to Do Things Your Way
How to Have a Winning Job Interview
How to Land a Better Job
How to Write a Winning Résumé
Joyce Lain Kennedy's Career Book
Life Plan
Planning Your Career Change
Planning Your Career of Tomorrow
Planning Your College Education
Planning Your Military Career
Planning Your Own Home Business
Planning Your Young Child's Education

SURVIVAL GUIDES

High School Survival Guide
College Survival Guide

VGM Career Horizons
a division of *NTC Publishing Group*
4255 West Touhy Avenue
Lincolnwood, Illinois 60646-1975